The Student Edition of
Working Model®
Version 2.0

Carol A. Rubin

▲▼ Addison-Wesley Publishing Company

Reading, Massachusetts · Menlo Park, California · New York · Don Mills, Ontario · Wokingham,
England · Amsterdam · Bonn · Sydney · Singapore · Tokyo · Madrid · San Juan · Milan · Paris

Acquisitions Editor: Stuart Johnson
Project Manager: Cindy Johnson, Publishing Services
Development Editor: Kathleen Habib, Publishing Services
Assistant Editors: Jennifer Albanese, Melissa Honig
Senior Production Supervisor: John Walker
Senior Marketing Manager: Dave Theisen
Manager, Educational Technology: Janet Drumm
Text Designer: Jean Hammond
Package Designer: Suzin Purney
Art Director: Peter Blaiwas
Illustrator: Rolin Graphics
Copyeditor: Pam Mayne
Technical Validator: Jeff Andrews
Technical Support: Jan MacKay
Compositor: Fog Press
Manufacturing Coordinator: Patrick Mescall

ISBN 0-201-43961-1

12345678910—CRW—9998979695

Preface

Working Model® is a revolutionary motion simulation package that allows you to quickly build and analyze dynamic mechanical systems on desktop computers. A powerful tool for analysis, simulation, and animation, Working Model has become an integral tool in the design process for engineers of all types. Its intuitive interface makes it easy to learn and use, while its powerful physics-based dynamics engine enables it to accurately calculate the behavior of highly complex dynamic models. Working Model makes it possible for engineers at all levels to test, redesign, and retest their designs before building a physical model.

The Student Edition of Working Model combines the Working Model Version 2.0 software with a Tutorial Manual written especially for students. Affordable enough for every engineering student to own, *The Student Edition of Working Model* will be an indispensable tool in courses throughout the engineering curriculum. Its Tutorial Manual can be used in conjunction with an introductory engineering course or for self-study. In both cases, it gives students the opportunity to use Working Model throughout their college years.

The Student Edition of Working Model brings you all the functions and features of the professional version of Working Model Version 2.0, with a design capacity suitable for most student projects. With *The Student Edition of Working Model*, you can build models with up to 20 mass objects, and an unlimited number of points, constraints, and meters. You can also view and run models of any size created with the professional version of Working Model; you are limited only by the amount of memory available on your machine.

The Student Edition of Working Model offers all the features that make the professional version of Working Model 2.0 such a powerful simulation tool, including:

- Intuitive point-and-click interface (consistent, easy to use, mouse-driven)
- Fine control over simulation parameters and accuracy settings
- Ability to define forces and properties with equations written in any unit system
- Complete set of constraints, including gears, spline-based slot joints, rods, actuators, springs, ropes, dampers, pulleys, motors, and more
- Meters with graphical or digital output, plus the ability to export results in ASCII format
- Ability to import CAD designs via DXF files into Working Model
- Compatibility with DDE and Apple® Events so simulations can be controlled by Microsoft® Excel or MATLAB®

The Student Edition of Working Model is available for Windows™ and Macintosh® computers, and is highly compatible across platforms. The Tutorial Manual includes instructions for both versions, and most models created on one platform can be ported seamlessly to the other.

Objectives

The Student Edition of Working Model was designed to meet the following objectives:

- To present Working Model in a problem-solving context relevant to a wide range of courses in the engineering curriculum
- To provide an introduction to Working Model suitable for self-study
- To support visualization of physical processes being studied in introductory dynamics courses
- To encourage students to use Working Model on their own as a tool for homework and project assignments

Features of the Tutorial Manual

To achieve these objectives, the Tutorial Manual includes:

- Tutorials that center on a typical engineering problem or design challenge
- Complete installation instructions and technical appendixes for the computer novice
- Step-by-step tutorials written for the beginning user
- Tutorial objectives keyed to principles taught in an introductory engineering dynamics course
- "Link to Dynamics" and "Link to Design" features that relate Working Model to dynamics and design principles
- Challenging and interesting end-of-tutorial exercises
- An ongoing exercise in the design of an automobile suspension system that integrates Working Model concepts into the design process
- Selected exercises from Bedford and Fowler's *Engineering Mechanics: Dynamics* (Addison-Wesley, 1995) at the end of each tutorial, which illustrate the utility of Working Model for testing and visualizing homework solutions

Organization of the Tutorial Manual

The Tutorial Manual for *The Student Edition of Working Model* proceeds in a logical fashion to guide the student from installing the software to working with all of its tools in a design environment.

Part 1, Getting Started, introduces the package and its contents, provides step-by-step installation instructions, and takes the student on a guided tour of the Working Model screen display, icon-based tool palette, menu bar, help facility, and mouse usage conventions.

Part 2, Tutorials, introduces Working Model in the context of engineering problem-solving and design. Tutorials 1 through 10 show students how to build models to illustrate and measure the behavior of common dynamic systems. Tutorials 11 through 15 then build on these basics by having students use Working Model to design systems to meet problem specifications.

Tutorials 1, 2, and 3 introduce basic concepts of setting up simulations in Working Model and use these concepts to model systems that illustrate projectile motion, relative motion, and Newton's second law of motion.

Tutorial 4 shows how Working Model can be used for the very large systems necessary to model planetary motion.

Tutorials 5 and 6 introduce constraints that limit motion and explore rolling contact and rotational inertia. With these tutorials, students have completed the basics and begin to learn more about the parameters controlling their simulations.

Tutorial 7 explores impulse and momentum and the concept of energy conservation.

Tutorials 8, 9, 10, and 11 introduce more complex models, such as shock absorbers and linkages, and show how to model noncontinuous conditions, how to export data from Working Model, and how to add pictures to simulations.

Tutorials 12, 13, 14, and 15 focus on the design process through the construction of models of a cam-follower, epicyclic gear train, internal combustion engine, and batting machine.

Part 3, Appendixes, provides additional reference information for the student. Appendix A contains technical information about the methods of approximation used by Working Model's dynamics engine and other tips, Appendix B contains a complete guide to the formula language used by Working Model, and Appendix C contains the details of importing and exporting DXF files and using DDE/Apple Events.

Acknowledgments

It is important to acknowledge the many individuals who contributed to the conceptualization and implementation of the exciting idea that became *The Student Edition of Working Model*, as it was a true team effort.

First and foremost, I want to thank Robin Redfield of Texas A&M University for working with me as the tutorials developed and for contributing such interesting end-of-tutorial exercises. His exercises challenge students to go beyond the tutorials and use Working Model to explore the world of mechanical systems.

Addison-Wesley and I thank the many teaching colleagues, too numerous to name, who participated in our telephone research and helped us shape the vision for the Tutorial Manual.

We are also grateful to the reviewers listed below for their advice and contributions to the development of the individual tutorials and for sticking with us through an ambitious schedule.

Janet Allen, *Georgia Institute of Technology*

Joseph Guarino, *Boise State University*

Kenneth Halliday, *Ohio University*

Charles Krousgrill, *Purdue University*

Keith Nisbett, *University of Missouri, Rolla*

Charles Procter, *University of Florida*

Robin Redfield, *Texas A&M University*

David Rosen, *Georgia Institute of Technology*

Andy Ruina, *Cornell University*

Steve Toebes, *Raytheon Corporation*

Student Reviewers:

Tanya Mamedalin, *Cornell University*

Sherilyn Kay Orr, *University of Idaho*

We also wish to thank David and Greg Baszucki, Peter Goettner, Umberto Milletti, and others at Knowledge Revolution who lent their talents to the partnership with Addison-Wesley. Their commitment to educators was instrumental in making this Student Edition a reality.

I would like to thank the people at Addison-Wesley who worked with me on this project, especially Jennifer Duggan, who so efficiently directed the distribution and evaluation of the tutorials and was solicitous of my every need. Thanks also to David Chelton, my very first contact with Addison-Wesley, who was extremely helpful with suggestions for improving the text.

The two people most responsible for producing this manual are Cindy Johnson and Kate Habib of Publishing Services. They expertly guided the editing, testing, and production of the step-by-step tutorials and worked diligently to assemble a manual that is complete, technically accurate, and pleasing to look at; and they did it with tact and grace. They introduced me to the publishing business at the start of this project, and guided me carefully and gently through the entire process, making it a pleasure rather than a task. I can't imagine a better technical editor than Kate, nor can I imagine two people with whom I would rather work.

Carol A. Rubin

Contents

Part Three ■ Appendixes

PART ONE

Getting Started

Before You Begin

Objectives

This chapter describes the contents of *The Student Edition of Working Model* package, defines the equipment you'll need to use the program, and previews the types of instructions you'll encounter in the chapters that follow. When you have finished reading this chapter, you'll be able to:

- Examine the contents of your Working Model® package
- Check your computer setup
- Recognize the typographical conventions used in the book
- Start the computer
- Work with diskettes
- Make backup copies of Working Model
- Copy Working Model onto a hard disk
- Start Working Model
- Exit Working Model

What Is Working Model?

Knowledge Revolution's *Student Edition of Working Model Version 2.0* is an advanced motion simulation package with sophisticated editing capabilities. It allows you to build and analyze dynamic mechanical systems, quickly and easily, on your computer. Working Model has become an integral tool in the design process for engineers of all types. With the tutorials in this manual, Working Model is easy to learn and easy to use. This software makes it possible for engineering students at all levels to test, redesign, and retest their designs before building a physical model.

The Student Edition of Working Model brings you all the functions and features of the professional version of Working Model Version 2.0, with a design capacity suitable for most student projects. With *The Student Edition of Working Model*, you can build models with up to 20 mass objects, and an unlimited number of points, constraints, and meters. You can also view and run models of any size created with the professional version of Working Model; you are limited only by the amount of memory available on your computer.

Examining the Contents of Your Working Model Package

This package combines all the software and information you need to use the Working Model program to create and run your own simulations on either a Windows™ or a Macintosh® system.

The Student Edition of Working Model contains instructional and reference material for using Working Model to create and run simulations. Your package should contain all of the following items:

The Tutorial Manual (this book)

Two 3.5-inch 1.44-MB disks

A License Agreement

A Warranty Registration Card

The Working Model tutorials in this Tutorial Manual contain hands-on instructions for developing increasingly complex simulations. Three appendixes provide additional technical and reference information that will help you do more with Working Model.

The installation disks contain the Working Model program, along with sample simulations created with Working Model and data files needed for the tutorials in this manual.

Please read the License Agreement, and read, complete, and mail the Warranty Registration Card immediately.

Combined format for Windows and Macintosh

Many of the illustrations used in this manual show either Windows or Macintosh dialog boxes. Both versions of a dialog box appear only when the two are substantially different. Any information pertaining to only one of the systems is labeled as such, with information specific to Windows set off by the **W** symbol and information specific to Macintosh set off by the **M** symbol.

Obtaining Product Support

Neither Addison-Wesley nor Knowledge Revolution provides telephone assistance to users of *The Student Edition of Working Model*.

However, registered instructors who have adopted this product for their students are entitled to telephone support. You should report any problems you encounter with the Working Model software to your instructor. Be sure to note exactly what action you are performing when the problem occurs, as well as the exact error message. Save your simulation, if possible, to a diskette so that your instructor can observe the problem.

Checking Your Computer Setup

W Windows systems

Working Model *requires* the following hardware and system software:

- A 386-based computer with a math co-processor or better (486DX is recommended) running Microsoft® Windows Version 3.1 or greater.

- 8 MB of RAM (Working Model will run with 4 MB, but you will have to turn on virtual memory. Using virtual memory has a significant negative impact on Working Model's performance). 16 MB of RAM is recommended for good performance.

- Approximately 8 MB of free disk space (to install Working Model and save simulations). 20 MB is recommended for good performance.

- A mouse and a display compatible with Microsoft Windows.

- A 1.44-MB 3.5-inch disk drive (for installation of Working Model).

Note: Occasionally Working Model may have insufficient system memory (RAM) at its disposal, especially if you are running with 4 MB of RAM, and it will warn you accordingly. To increase the amount of memory available to Working Model (and the other Windows applications), you should turn on virtual memory. To do so, run the 386 Enhanced application in the Windows Control Panel and turn on Virtual Memory. There will be a performance tradeoff.

Device drivers

Working Model supports the standard Windows device drivers provided by Microsoft. Working Model may not work properly with device drivers provided by other manufacturers.

M Macintosh systems

- Working Model runs on Macintosh computers using System 7 or greater (Mac II or above). Working Model relies on Color Quick Draw, which is supported only by Macintoshes with 68020s or higher. In particular, the Macintosh Plus, SE, Portable, Classic, and Powerbook 100 are not supported. Working Model is compatible with Power PC-based Macintosh computers.

- The memory requirement is affected by the number of colors you are using as well as the number of documents you want to leave open at one time. Use the following guidelines to determine your requirements.

1. To run Working Model on a *black and white system* and keep *only one document open,* you need approximately 2.5 MB of free memory. (This typically means your machine needs a total of 4 MB of physical RAM, considering that System 7 takes up more than 1 MB.)

2. To run Working Model in *256 colors* and keep *only one document open,* you need approximately 3.5 MB of free RAM.

3. If you want to run Working Model in *256 colors* and keep *two documents open,* you need approximately 4 MB of free RAM.

Note: On the Macintosh, you should turn on virtual memory to increase the amount of memory available to Working Model. Use the Memory control panel to turn virtual memory on. There will be a performance tradeoff.

FPU support

Working Model is designed to take full advantage of the Floating Point Unit (math coprocessor), if one is available on your Macintosh computer. (However, it is not required to run Working Model.) An FPU speeds the computations enormously, allowing for faster, smoother animation.

Recognizing Typographical Conventions

Mouse terminology

You use the *mouse*—a kind of pointing device—to begin most operations in Windows and on the Macintosh. When you move the mouse on your desk, a corresponding *pointer,* such as an arrow, makes the same movement on the screen. (The pointer is not always an arrow—it often takes a different form, such as an I-beam for editing text, or a crosshair for selecting graphics.)

Table GS1.1 describes some of the terms referred to throughout this manual. Other terms are explained as they are introduced in the text.

W Note: The mice used with Microsoft Windows have more than one button. Use the left mouse button, which is the default. You can change the default with the Windows Control Panel.

Choosing and selecting

Choose and *select* are used interchangeably throughout the tutorials. They indicate that you should click or double-click on a tool, menu item, or Working Model mass object with your mouse.

The tool, menu item, or Working Model mass object that you are to choose or select is set in **boldface**.

Windows vs. Macintosh

Working Model operates almost identically on the Windows and Macintosh platforms, and allows you to move files back and forth between the two platforms. In some cases, however, commands or dialog boxes are slightly different. Where they are, Windows instructions are set off with a **W** symbol. Macintosh instructions are set off with a **M** symbol.

New Working Model terms

When a Working Model term is being introduced and explained for the first time, it is *italicized.*

Table GS I.I Mouse Terminology

Term	Meaning
Point	*Pointing* is simply moving the pointer onto an object or area on the screen.
	To point, move the mouse until the mouse pointer on the screen touches the item you want.
	If you run out of room on your desk, you can pick up the mouse and place it at another spot. The pointer on the screen is controlled by the motion of the ball on the bottom of the mouse and will not move while the mouse is off the table top.
Click	*Clicking* begins an action at the pointer's location.
	To click, quickly press and release the mouse button without moving the mouse.
Double-click	*Double-clicking* is usually a way to display the contents of an icon or document, or to start an application program.
	To double-click, click the mouse button twice in rapid succession without moving the mouse.
Shift-click	*Shift-clicking* is a variation of clicking that you use to select multiple icons or other objects that are not near each other, or to select only some of the items in a group.
	To shift-click, press the Shift key on the keyboard and click on each object that you want to select.
	Each object becomes *highlighted* when you click on it; that is, it changes color.
Hold down	*Holding down*, or *holding*, keeps the icon highlighted (selected), or continues to show the items on a menu so you can view them.
	To hold down, press the mouse button without releasing or moving the mouse.
Drag	*Dragging* allows you to move an item on the screen as you move the mouse. You also drag to move an icon or a window and to select a group of icons.
	To drag, press and hold down the mouse button while you move the mouse.
	The object remains highlighted when you release the button.

Starting Your Computer

To install and use Working Model, you must have a computer with a hard disk. A hard disk is a fixed, permanent disk that can store the contents of many floppy disks.

W If you're working with a hard disk that has DOS installed on it, begin by turning on the power and, if necessary, the monitor.

W If your system asks you to enter the date and time, do so and wait for the operating system prompt (usually C:\>) to appear.

ℳ Begin by turning on the power and, if necessary, the monitor.

Working with Diskettes

The Student Edition of Working Model comes on diskettes in the 3.5-inch format. Even though you have a hard disk, you'll be using the diskettes to load the program and to store copies of the simulations you create.

Taking care of disks

Whenever you work with disks, keep the following precautions in mind:

- Never touch the exposed areas of a disk, and never handle a 3.5-inch disk with its shutter open.

- Keep any disk away from heat, sunlight, smoke, and magnetic fields (telephones, televisions, transformers, monitors, etc.).

- If the disk drive's slot is horizontal, slide the disk into the drive with the label facing up. If the drive is mounted vertically, the label should face left. Insert the metal shutter first.

- Insert the disk into the drive as far as it will go, but don't force it.

- Never remove a disk while the drive access light is on.

Write-protecting disks

In addition to physical dangers, the information stored on disks is vulnerable to human error. Whenever you're working with an original disk, you can avoid accidents by write-protecting it. Write-protecting prevents you from accidentally changing or erasing any information on the disk. However, write-protecting does not prevent you from running a program from the disk or copying information from it.

The procedure for write-protecting is very simple. There is a small write-protect window in the upper-right corner of the disk. A plastic tab on the back of the disk slides up and down to open and close the window. When the window is open, the disk is write-protected. To prevent accidental erasure, your *Student Edition of Working Model* disks have been write-protected for you.

- If your program disks are not write-protected, slide the small plastic tab up to open the window.

Backing Up Your Working Model Disks

Before installing Working Model, please make backup copies of the installation disks. A backup is a copy of an original disk. We don't recommend using the original disks to install Working Model. Always work with backups of Working Model (or any software), and store your originals in a safe place. That way, you can always make another copy from the original if anything happens to the backup copy you're using.

ℳ Backup procedure for Windows

If you are working with the Macintosh, please skip to page GS-9.

Your computer should be running Windows. The steps below assume you have one 3.5-inch disk drive.

- If necessary, type the command (or command combination) to start Microsoft

Windows (generally, you would type
cd\windows, press the Enter key, type **win**,
and press the Enter key again).

- Insert Disk 1 in your disk drive.

- Double-click on the Main Program group
  icon to open it.

- Double-click on the File Manager icon to
  open it.

- Select **Disk** menu, **Copy Disk...**

The Copy Disk dialog box opens. The defaults
for Source and Destination should both be set
to A:

> **Note:** These instructions assume that your
> disk drive is drive A: If the name of your
> drive is different, please substitute the
> name of your drive for **A:** in the following
> instructions.

- Select **OK**.

When you finish, the following warning message appears: *This operation will erase ALL
data from the destination disk. Are you sure you
want to continue?*

- Select **Yes**. (If you select **No**, no backup
  takes place).

The following prompt should appear: *Insert
source disk.*

- At this prompt, insert the original Working
  Model Disk 1 and select **OK**.

When Windows has copied the contents of the
disk, the following message appears: *Insert
destination disk.*

- Remove Working Model Disk 1 and insert a
  blank formatted disk.

- Select **OK**.

Your system may not be able to copy the entire
contents of the source disk into memory in
one operation, and you may have to insert the
source disk again. Just follow the on-screen
instructions for switching disks in the drive.
Remember that the original Working Model
disk is the source disk, while the blank disk
that you are using to create a backup is the
destination disk.

- Repeat the copying process just described to
  copy Disk 2.

When you have finished, remember to write
the disk's title (for example, SE Working
Model Disk 1 Backup) on the label.

You are now ready to install Working Model
on your hard disk; please skip to page GS-10.

Ⓜ Backup procedure for Macintosh

If you are working in Windows, please return
to page GS-8.

Your computer should be on and the Finder
should be running.

- Insert the original Working Model Disk 1
  into the disk drive.

- Select **Special** menu, **Eject Disk**.

- Insert a blank disk into the disk drive.

A message window may appear that reads:
*This disk is unreadable Do you want to
Initialize it?* If this disk has been used with
another type of computer, the message tells
you that the disk is not a Macintosh disk.

- Select **Initialize**.

Another message window may appear, reading
*This process will erase all information on this
disk.*

- Select **Erase** to continue initialization.

Another message window appears, prompting you to name this disk.

- Name your disk **Disk 1 Backup**
- Select **OK**.
- Select **Special** menu, **Eject disk**.
- Click on the icon of the original Working Model Disk 1 and drag to the icon of Disk 1 Backup.
- A warning window appears on your screen, reading *Are you sure you want to completely replace contents of "Disk 1 Backup" (internal) with contents of "Disk 1" (not in any drive)?*
- Read the message to make sure you're making the correct decision, then choose **OK** to begin the copying process. (If you choose **Cancel** rather than **OK**, no backup takes place.)

Your system may not be able to copy the entire contents of the source disk into memory in one operation, and you may have to insert the source disk again. Just follow the on-screen instructions for switching disks in the drive. Remember that the original Working Model disk is Disk 1, while the blank disk that you are using to create a backup is Disk 1 Backup.

- Follow the instructions provided for switching the disks as needed.
- When you have completed the copying process, drag the icons of both disks to the trash to eject them.
- Repeat the copying process just described to make a backup copy of Disk 2.

When you have finished, remember to write the disk's title (for example, SE Working Model Disk 1 Backup) on the label.

You are now ready to install Working Model on your hard disk; please skip to page GS-12.

Installing Working Model on a Hard Disk

This section includes instructions for installing Working Model and the demonstration files on a hard disk. Working Model comes in a compressed format. The installation program uncompresses the Working Model files and installs them on your hard disk.

W Windows installation

If you are installing on the Macintosh, please skip to page GS-12.

Viewing the README.TXT file

Before you install Working Model, please read the text file named README.TXT or README.WRI, located on Disk 1. In this file you will find useful information about installing and using *The Student Edition of Working Model.* You can open and print the file in any word processor, or read and print it by double-clicking on the file name in the Windows File Manager.

Installing the Working Model program

To install the Working Model program and files on your Windows system:

- Type the necessary command (or command combination) to start Microsoft Windows (generally, you would type **cd\windows**, press the Enter key, type **win**, and press the Enter key again).
- Insert Disk 1 in your disk drive.

- Choose **File** menu, **Run...** as shown in Figure GS1.1.

Figure GS1.1

- Type **a:\setup** in the text box, as shown in Figure GS1.2.

  Note: These instructions assume that your disk drive is drive A:. If the name of your drive is different, please substitute the name of your drive for **A:** in the following instructions.

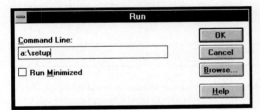

Figure GS1.2

- Choose **OK**.

The Installation program starts.

- Respond by clicking **Continue** or **OK** in response to each of the instruction screens that appear to continue with your installation.

The Installation Options dialog box shown in Figure GS1.3 appears.

Figure GS1.3

- If you want to install all of Working Model, choose **Standard Installation**; otherwise, select **Custom Installation**.

  Note: If you choose Custom Installation, you can decide which files to install. The Working Model *application files* are the Working Model program files. You must install these files to run Working Model. The *sample files* are pre-built simulations that show you some of the problems you can solve with Working Model. These files are optional. The *data files* are data sets needed for one of the tutorials in this manual. You must install these files to complete Tutorial 12.

Working Model automatically sets up a default directory, WM2.0S, for your application and files. You can change the name and location of this file if you wish.

- Select the default directory or rename the directory where you want Working Model to be installed.

- Select **Continue**.

Note: In the Custom Installation dialog box, you need to click the Set Location button to change the directory.

Note: You may get a message window that says: *Insufficient Disk Space. You do not have enough disk space to install all of the files you have selected. Please remove some files from your set of selections or change your installation destination.* If you get this message and you have selected only the program files, be sure you have indicated the correct hard disk drive for the installation. If you have, try to move some old files from your hard disk to diskettes to make room for Working Model. When you have made more space available (at least 8 MB), try the installation process again.

- Insert Disk 2 when prompted.
- Select **OK**.

A final Exit screen reports that your setup succeeded.

- Select **OK**.

The Student Edition of Working Model for Windows is now installed, and its icon is placed in the Program Group Working Model, and called Working Model Student Edition. Be sure to store the original disks in a safe place!

You are now ready to learn how to load and start *The Student Edition of Working Model* on your computer system. Please skip to Starting Working Model on page GS-14.

Note: If you Exit in the middle of the installation, some files will be left in the WM2.0S directory on your hard drive. You will need to delete them with DOS or Windows commands.

Ⓜ Macintosh installation

If you are installing Working Model in Windows, please return to page GS-10.

Installing the Working Model program
You should disable all virus protection before attempting to install Working Model, as it may interfere with the installation procedure. To install the Working Model program and files on your Macintosh system:

- If necessary, turn off all virus protection.
- Insert Disk 1 in the floppy drive of your Macintosh.

A folder containing an icon labeled "Double-click me to install" opens on your desktop as shown in Figure GS1.4.

- Double-click on the **Double-click me to install** icon.

Figure GS1.4

- When the first instruction screen appears, click **OK** to continue with your installation.

Viewing the Read Me file

A second instruction screen appears. This is the Read Me text included with this software.

Before you install Working Model, please read this. In this file you will find useful information about installing and using *The Student Edition of Working Model.* You can save or print the file from the Installer.

- Click **Continue** to continue with your installation.

A dialog box appears with various installation options, as shown in Figure GS1.5.

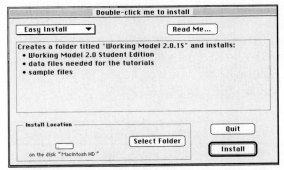

Figure GS1.5

Floating Point Unit vs. non-FPU installation

There are two different versions of Working Model on your installation disks. One of them is designed to make optimal use of an FPU if it is available.

The FPU version of Working Model compiles floating point arithmetic directly into FPU instructions. This version *requires* the presence of a coprocessor and provides significant gains in performance over the non-FPU version, because Working Model is a floating point-intensive application.

The installation software automatically determines whether a math coprocessor is present and installs the correct version of Working Model.

PowerPC users

If you have a Macintosh PowerPC™ and choose the Custom Install option, you *must* install the non-FPU version of Working Model. The PowerPC performs fast floating point operation without specific FPU instructions.

- If you want to install Working Model and the demonstration files on your hard disk, leave the pull-down menu at the upper-left side of the dialog box set to **Easy Install** and choose the **Install** button. Otherwise, select **Custom Install** from the pull-down menu to specify a non-standard combination of Working Model files for installation and choose the **Install** button.

Note: If you choose Custom Install, you can decide which files to install or to install the non-FPU version of Working Model even though an FPU is installed on your computer. The Working Model *application files* are the Working Model program files. You must install these files to run Working Model. The *sample files* are pre-built simulations that show you some of the problems you can solve with Working Model. These files are optional. The *data files* are data sets needed for one of the tutorials in this manual. You must install these files to complete Tutorial 12.

By selecting the items you want to install and by using the Select Folder options, you can install as much of the Working Model package as you want, in whatever location is convenient for you.

The installer automatically detects whether an FPU is available on your machine and installs the correct version of Working Model. All of the data and demonstration files are also installed. Working Model automatically sets up a default directory, Working Model 2.0.1S, for your application and files. You can change the name and location of this directory if you wish.

- Click the **Switch Disk** or **Select Folder** button to install on a different hard disk or in a different folder, if you wish.

Note: Switch Disk will only appear if you have more than one hard disk available, otherwise Select Folder will appear. If Switch Disk appears, Select Folder is an option on this pull-down menu.

The installer prompts you for an alternate hard disk on which to install.

- Select the name of the appropriate hard disk and folder.
- Select **OK**.
- Insert Disk 2 as you are prompted for it.

Note: Depending on the configuration of your Macintosh, you may not be prompted for Disk 2 if you choose Easy Install.

When you have finished the installation, you are prompted to install another copy of Working Model, if you wish. Follow the instructions on the screen until you are done.

The Student Edition of Working Model for the Macintosh is now installed, and its icon is placed on your hard drive and called Working Model SE (you can move the icon to another folder if you prefer). Remember to turn your virus protection back on. Be sure to store the original disks in a safe place!

You are now ready to learn how to load and start *The Student Edition of Working Model* on your computer system.

Starting Working Model

The following instructions assume that you have installed Working Model on your hard disk as described in the preceding sections.

Your computer should be on, with the **W** Program Manager or the **M** Finder™ on your screen. Open the **W** Working Model program group or **M** Working Model 2.0.1S folder, so that the **W** Working Model Student Edition or **M** Working Model SE icon is visible, as shown in Figure GS1.6.

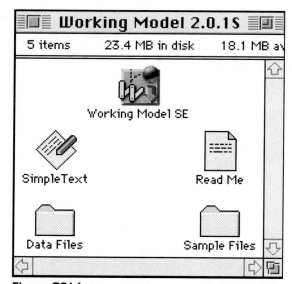

Figure GS1.6

The quickest way to open an application from its icon is to double-click on the icon.

- Double-click on the Working Model (**w** Student Edition or **M** SE) icon to start the program.

   Tip Another way to open an application from its icon is by selecting the icon (clicking on it once) and choosing the Open command from the File menu.

A dialog box like the one in Figure GS1.7 appears on your screen, asking you for your name, the name of your organization, and the serial number for your software.

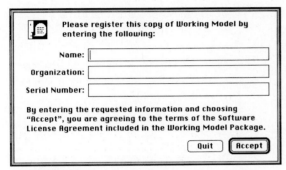

Figure GS1.7

- Enter your name.
- Press the **Tab** key.
- Enter the name of your school.
- Press the **Tab** key.

   Note: The Working Model Serial Number dialog box is not case sensitive; you can type any letters in upper case or lower case. You must include all spaces and dashes exactly as they appear on your disk when entering your serial number, or it will not be accepted.

- Carefully enter the serial number from your Working Model Disk 1.
- Click **w OK** or **M Accept**.

A dialog box like the one in Figure GS1.8 appears on your screen, asking for the 3-digit code on the bottom of a page from this manual. This code appears on the inside margin of each page and should be entered in the order in which it appears. This is a copy protection code that you will be asked for each time you start *The Student Edition of Working Model*. You will want to keep your manual handy for each session with Working Model.

Figure GS1.8

- Enter the requested code.
- Click **OK**.

If you make a mistake, or the code is not accepted, Working Model prompts you again. After the third try, Working Model asks you to try again at another time. If this happens, open the program again from its icon and enter the requested code.

The Student Edition of Working Model splash screen appears, identifying you as the registered licensee of the software. Then Working Model starts up and opens a new, untitled document, as shown in Figure GS1.9.

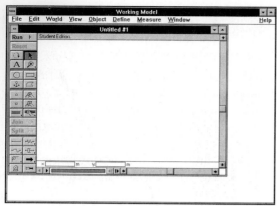

Figure GS 1.9

Exiting Working Model

To exit Working Model:

- Select **File** menu, **w** **Exit** or **M** **Quit**.

A message window may appear on your screen, asking if you want to save your file.

- Select **w** **No** or **M** **Don't Save**.

Working Model quits to **T** Microsoft Windows or **M** the Finder. You have successfully installed, loaded, and exited Working Model. Chapter 2 will introduce Working Model's menus and features.

Working Model Basics

Objectives

When you have completed this chapter, you will be able to:

- Start Working Model
- Work with a window
- Recognize Working Model tool bar and menu command options
- Use keyboard functions in Working Model
- Understand the Smart Editor
- Understand Working Model's methods of numerical integration
- Exit Working Model

This chapter teaches you how to use Working Model's menus and dialog boxes as you are introduced to the basics of a Working Model simulation. When you have finished this chapter, you will be ready to complete the tutorials in this manual.

Starting Working Model

If you have not installed Working Model on your hard disk, please refer to Chapter 1 for installation instructions.

Your computer should be on, with **W** the Program Manager or **M** the Finder on your screen. Open the **W** Working Model program group or **M** Working Model 2.0.1S folder, so that the **W** Working Model Student Edition or **M** Working Model SE icon is visible, as shown in Figure GS2.1.

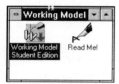

Figure GS2.1

The quickest way to launch a program from its icon is to double-click on the icon.

- Double-click on the Working Model (**W** Student Edition or **M** SE) icon to start the program.

A dialog box appears on your screen, asking for the 3-digit code on the bottom of a page from this manual. This code appears on the inside margin of each page and should be entered in the order in which it appears. This is a copy protection code that you will be asked for each time you start *The Student*

Edition of Working Model. You will want to keep your manual handy for each session with Working Model.

- Enter the requested code.

- Click **OK**.

The Student Edition of Working Model splash screen appears on your screen, identifying you as the registered licensee of the software.

The new, untitled simulation document appears in its own window. You will see the tool bar on the left and the Tape Player controls along the bottom of the window. The document window may not open so that the entire window is visible, or it may not fill the entire screen. To maximize your window to fill the screen, click on the **W** Maximize arrow or **M** Zoom box in the upper-right corner of the document. More details on working with windows and a mouse are given below. If you are an experienced Windows or Macintosh user, you do not need to review these basics; please skip to The Working Model Cursor on page GS-21.

Working with a Window

When you work with applications, you often have a number of windows open. This section explains how to move and resize windows, scroll through their contents, and arrange them on the desktop.

Resizing a window

You can change the size and shape of a window so that you can view the contents of two or more windows simultaneously. To change the size of a window:

W Point to a border or corner of the window. The pointer changes to a two-headed arrow.

W Drag the corner or border until the window is the size you want. If you drag a corner, the two sides that form the corner are sized at the same time.

M Place the pointer in the size box in the window's lower-right corner.

M Drag; an outline of the window shows the new size.

■ Release the mouse button.

Enlarging a window

In most cases, you can enlarge a window to fill a larger portion of the desktop, or even the entire desktop. To enlarge a window to its maximum size:

W Click the Maximize button (the up arrow) in the upper-right corner of the window. The Maximize button is replaced by the Restore button.

In Windows, a maximized document window shares the main window's title bar. The Minimize and Maximize buttons affect the application window, not any window inside it. Only the Restore button affects the document window alone.

W Another way to maximize a window is by double-clicking on its title bar. To restore an application window to its previous size, double-click on the title bar again.

M Click the Zoom button in the upper-right corner of the window.

Moving a window

You can move any window to a different location on your desktop. To move a window, place the pointer in the window's title bar at the top of the window. Drag the window to a new location. An outline of the window follows the pointer. Release the mouse button. (This will not work with a maximized window in Windows.)

Closing a window

To close a window, use one of the following methods:

■ Choose **File** menu, **Close**
or Press **W** **CTRL+W** or **M** **Command+W**;
or **W** Choose **Control** menu, **Close**;
or **W** Double-click on the Control menu box in the upper-left corner;
or **M** Click on the close box in the upper-left corner.

Switching between windows

You make an open window active by clicking inside it. You may have to move or resize other windows if the one you want is open but covered by other windows.

To make a window active, click anywhere in that window. When you click, the window moves in front of the other items displayed on the screen. **M** Another way to make a window active is to click on its dimmed icon.

W Reducing a window to an icon in Windows

You may want to reduce an application window to an icon if you want the application available for later use. When reduced to icons, Windows-based applications continue to run, but their windows do not take up space on the desktop. After a window is reduced to an icon, you can select and move the icon in the same way you select and move a window.

To reduce a window to an icon, select the window you want to reduce. Click the Minimize button (the down arrow) in the upper-right corner of the window.

When a document window is maximized, the Minimize and Maximize buttons are replaced by the Restore button. If you want to minimize the document, you can either choose Minimize from the window's Control menu, or you can first restore the window and then click the Minimize button.

Restoring an icon or window

You can restore an icon or window to its original size.

W To restore an icon to a window, double-click on the icon. To restore an enlarged window to its previous size, **W** click the Restore button in the upper-right corner of the window, or **M** click the Zoom button in the upper-right corner of the window.

Viewing the contents of a window

You can see a window's contents in two ways—by adjusting the size of the window or by *scrolling* to make a different part of the window visible. You have to scroll to see the contents of documents that are longer or wider than the screen.

Using scroll bars

Some windows and dialog boxes have scroll bars that you can use to view information that does not fit inside the window.

The amount you see in the window at any one time depends on the window's size. Every window has two scroll bars, one for vertical scrolling and one for horizontal. For both, a gray bar indicates more content beyond the

window's borders. **W** When you can view all the contents of a window or dialog box without scrolling, the scroll bars may be absent or dimmed to indicate they cannot be used. **M** A white, or clear, bar indicates that all the content is visible.

The up arrow displays content above the window's top border; the down arrow displays content below the window's bottom border. The left and right arrows display content beyond the left and right borders of the window, respectively. To scroll through the contents of a window, place the pointer on one of the scroll arrows and hold down the mouse button. The window's contents move to show the part that was hidden. The scroll box on the scroll bar also moves in the direction of the arrows; you can grab the box and move it up or down, or left or right. Release the mouse button when the item you want to see is visible. Another way to use the scroll arrows is simply to click on a scroll arrow to scroll in small increments.

To scroll through all the information displayed in a window or dialog box, drag the scroll box.

Mouse Basics in Working Model

You use the *mouse*—a pointing device—to begin most operations in Windows and on the Macintosh. When you move the mouse on your desk, a corresponding *pointer*, such as an arrow, makes the same movement on the screen. See Mouse terminology in Chapter 1 for details on using the mouse in Working Model.

You are now ready to begin a guided tour of the Working Model workspace.

The Working Model Cursor

The Arrow tool is the default cursor in Working Model. It is used for picking menu selections and tools, and for selecting objects.

Clicking on an icon on the tool bar changes the visual appearance of the cursor, sometimes to look like the icon for that tool. This is a cue to what tool you have active in Working Model.

The Working Model Workspace

The tool bar contains tools you will use to create simulations. Tools are provided for creating masses, springs, ropes, forces, and many other objects. The tool bar also contains buttons for running and resetting simulations.

The Tape Player Controls give you more options for running and viewing simulations. You can use the controls to step through a simulation backwards (as you would play a tape in reverse), or to move to a specific frame in the simulation.

The Help Ribbon gives a concise description of the tool or object located at the mouse cursor.

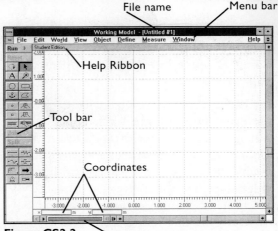

Figure GS2.2

The Working Model Tool Bar

Every Working Model simulation window has a tool bar on the left side, as shown in Figure GS2.2. The tool bar provides point-and-click access to the tools you will use to create simulations.

The tool bar contains a series of boxes, each of which displays an icon that represents a specific tool. Some of these boxes are pop-up palettes (to save space and to provide easy access to a set of tools). A pop-up palette is indicated by a small arrow in the lower-right corner of the box.

To open a pop-up palette, position the pointer on the arrow, and hold down the mouse button. The pop-up palette appears to one side of the box. Drag the pointer to highlight the desired tool and release the mouse button.

The tool selected last is displayed in place of the previous one. You can then click on the icon to re-select the tool without opening the pop-up palette again.

If you click once on a tool, it will be selected for the next operation; after that operation the cursor reverts to the Arrow tool. To use the tool for several successive operations, you can double-click on it and it will remain active indefinitely. The difference (one-time use vs. indefinite activation) is indicated on the tool palette by the color of the icon's box: a double-clicked item is dark gray, while a single-clicked item is light gray. To quickly select the Arrow tool, press the space bar.

Starting from the top, and moving left to right, the Working Model tool bar consists of the following tools: Run, Reset, Rotate, Arrow, Text, Zoom, Circle, Rectangle, Anchor, Polygon, Point Element, Pin Joint, Square Point Element, Rigid Joint, Slot Element, Slot Joint, Join, Split, Rod, Spring, Rope, Damper, Pulley System, Force, Motor, and Actuator. See Figure GS2.3.

Figure GS2.3

Run

Use this button to start a simulation. When a simulation is in progress, the Run button is replaced by a Stop button. Clicking the Stop button or in the workspace stops the simulation. You can resume the action of the simulation from its last frame (i.e., where you stopped it) by clicking the Run button again.

When the simulation is running, "Running" appears in the Help Ribbon.

Reset

Use this button to stop your simulation (if it is still running) and bring the simulation back to its initial conditions and its first frame.

Rotate

Use this tool to rotate an object or a selected group of objects. You can rotate objects about their center of mass, as well as about pin joints and measurement points. You will see a line snap to the points about which objects can be rotated.

Arrow

Use this tool to select an object or group of objects, or to drag a selected group of objects on the screen. The Arrow tool is the Working Model default; when you open a new window, the Arrow tool is active.

Text

Use this tool to enter text directly into the simulation workspace.

Zoom

The Zoom pop-up palette has two selections: Zoom In and Zoom Out.

Zoom In

This tool increases the magnification of the workspace by a factor of two (2x) when you click in the workspace. The new view is centered on the area around the pointer. Holding down the Shift key while you click in the workspace toggles this tool to the Zoom Out tool.

Zoom Out

This tool decreases the magnification of the workspace by a factor of 2 (½x). Holding down the Shift key while you click in the workspace toggles this tool to the Zoom In tool.

The Circle, Rectangle, Square, and Polygon tools create mass objects in the workspace.

Circle

Use this tool to create circular mass objects.

- Click on the **Circle** icon on the tool bar.
- Click once in the workspace to create a circle.
- Double-click on the **Circle** icon on the tool bar.
- Click twice in the workspace to create two more circles.
- Click on the **Arrow** icon to deselect the Circle tool.

Your workspace should look like Figure GS2.4.

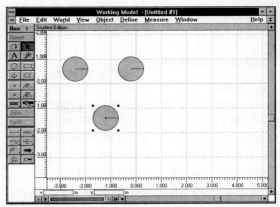

Figure GS2.4

Rectangle

The Rectangle pop-up palette has two selections: Rectangle and Square.

Rectangle

Use this tool to create rectangular mass objects. You can also create a square with the Rectangle tool.

Square

Use this tool to create square mass objects. You cannot create a rectangle with the Square tool.

Anchor

Use this tool to lock the motion of mass objects. Anchored masses will not move unless you enter an equation in their Properties window to control their position or velocity.

Polygon

Use this tool to create irregularly shaped mass objects. Define each point with a single click. Double-click to signal the last point; the polygon automatically closes, connecting the first point with the last point.

▣ Point Element

Use this tool to create a point. You can use points for measuring properties at a specific location on a mass object. You can combine a point with a slot or another point to form a joint (a slot joint or a pin joint, respectively). Points have no mass, only location.

- Double-click on the **Point Element** icon on the tool bar.
- Click once each on two of the circles.
- Click on the **Arrow** tool to deselect the Point Element tool.

Your workspace should look like Figure GS2.5.

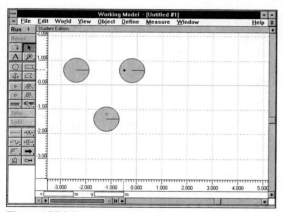

Figure GS2.5

The Pin Joint, Rigid Joint, Slot Element, Slot Joint, Join, Split, Rod, Spring, Rope, Damper, Pulley System, Force, Motor, and Actuator tools create constraints.

▣ Pin Joint

Use this tool to create a pin joint. A pin joint prevents separation between two mass objects, or one mass and the background at the location of the pin, but allows rotation.

▣ Square Point Element

Use this tool to create a square point element. You can combine a square point with a slot or another square point to form a joint (a keyed slot or a rigid joint, respectively).

▣ Rigid Joint

Use this tool to create a rigid joint. A rigid joint prevents movement or rotation between two mass objects, or between one object and the background.

Slot Element

The Slot Element pop-up palette has four selections: Horizontal Slot, Vertical Slot, Curved Slot, and Closed Curve Slot. A slot joined to a point forms a slot joint. A slot joined to a square point forms a keyed slot joint. The slot extends beyond the boundaries of your workspace.

▣ Horizontal Slot

Use this tool to create a horizontal slot element.

▣ Vertical Slot

Use this tool to create a vertical slot element.

▣ Curved Slot

Use this tool to create a curved slot from a series of smoothly interpolated (splined) control points. Define each control point with a single click, and double-click on the last point.

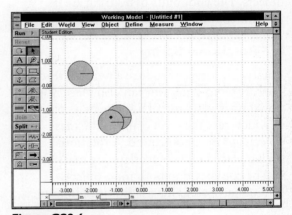 Closed Curve Slot

Use this tool to create a slot consisting of a closed curve. Define each control point of the curve with a single click. Double-click to signal the last point. The curve automatically closes, connecting the first point with the last point.

Slot Joint

The Slot Joint pop-up palette has five selections: Horizontal Pinned Slot Joint, Vertical Pinned Slot Joint, Horizontal Keyed Slot Joint, Vertical Keyed Slot Joint, and Curved Slot Joint.

Pinned Slot Joint

A pinned slot joint—horizontal, vertical, or curved—creates a point on one mass object and forces it to align with a slot on the second mass object or the background, while permitting rotation.

Keyed Slot Joint

A keyed slot joint—horizontal or vertical—creates a point on one mass object and forces it to align with a slot on the second mass object or the background, while prohibiting rotation.

Join

Use the Join button, shown in Figure GS2.6, to form a joint from two elements. Select both elements and then click the Join button. The elements need not overlap. Joining two point elements forms a pin joint. Joining two square point elements forms a rigid joint. Joining a slot element with a point forms a

pinned slot joint. Joining a slot element with a square point forms a keyed slot joint. The Join button also recombines elements that have been separated with the Split button.

- Click on the point on one circle in the workspace.
- Holding down the **Shift** key, click on the point on the other circle.
- Select the **Join** button on the tool bar.

Observe that the two circles have moved together in your workspace, and joined the points to form a pin joint, as shown in Figure GS2.6.

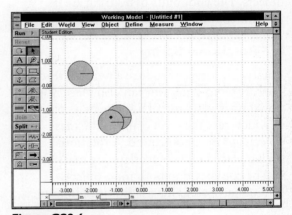

Figure GS2.6

Split

Use the Split button to temporarily separate a joint into its component elements. A dashed line indicates the separated elements of a constraint. The Join button recombines elements that have been separated with the Split button.

Rod

Use the Rod tool to create a massless, inflexible link between two mass objects. Rods cannot be compressed or extended. You can attach rods between one mass object and the background, or between two mass objects. The endpoints of a rod are its attachment points.

Spring

The Spring pop-up palette has three selections: Spring, Rotational Spring, and Spring-Damper.

Spring

Springs resist stretching or compression. You can attach springs between one mass object and the background, or between two mass objects. The endpoints of a spring are its attachment points.

Rotational Spring

Rotational springs produce a twisting force as they are wound up. You can also place rotational springs on top of a single mass object, in which case they will connect the mass object and the background. A rotational spring placed on two overlapping mass objects will be connected to both mass objects. Rotational springs have a built-in pin joint.

Spring-Damper

Spring-dampers provide a combination spring and damper. You can attach spring-dampers between a mass object and the background, or between two mass objects. The endpoints of a spring damper are its attachment points.

Rope

The Rope pop-up palette has two selections: Rope and Separator.

Rope

Ropes prevent objects from separating by more than a specific distance. Ropes go slack (and have no effect) when the objects they are connected to move close together. You can attach ropes between one mass object and the background, or between two mass objects. The endpoints of a rope are its attachment points.

Separator

Separators prevent objects from moving closer than a specific distance together. Separators have no effect when the objects they are connected to move far apart. You can attach separators between one mass object and the background, or between two mass objects. The endpoints of a separator are its attachment points.

Damper

The Damper pop-up palette has two selections: Damper and Rotational Damper.

Damper

Dampers resist changes in compression or extension. You can attach dampers between one mass object and the background, or between two mass objects. Dampers exert force proportional to the relative velocity between their two endpoints, in the direction of the line of action. The endpoints of a damper are its attachment points.

Rotational Damper

Rotational dampers produce a twisting force that resists changes in rotation. You can also place a rotational damper on top of a single mass object, in which case it will connect the mass object and the background. A rotational damper placed on two overlapping mass objects will be connected to both mass objects. Rotational dampers have a built-in pin joint.

Pulley System

The Pulley System pop-up palette has two selections: Pulley System and Gear.

Pulley System

Use the Pulley System tool to create a rope composed of several pulleys. Define each pulley with a single click. Double-click to signal the endpoint of the system. You can attach any pulley within a link to either the background or a mass object.

Since Working Model approximates pulleys as thread holes through which a rope is routed, they are massless and dimensionless.

Gear

Use the Gear tool to connect any two mass objects to act as a pair of rotational gears. Click on two objects to define a pair of gears (the objects need not be touching one another). The Gear tool creates a pair of external gears by default; you can use the Gear Properties window to make one gear an internal gear.

Force

The Force pop-up palette has two selections: Force and Torque.

Force

Forces act on the mass object to which they are attached. You can position the point of application anywhere on the mass object. You can fix the direction of the force with respect to either the background, or the mass object.

Torque

Torques apply a twisting force. You can position the point of application anywhere on the mass object.

Motor

Motors exert a twisting force between two masses. Each motor has a built-in pin joint. You can place a motor on top of a single mass object, in which case it will connect the mass object and the background. A motor placed on two overlapping mass objects will be connected to both mass objects.

Actuator

Use this tool to create a constraint that exerts a force—torque, rotation, velocity, or acceleration—between its endpoints. You can attach actuators to two mass objects, or to one mass object and the background. The endpoints of an actuator are its attachment points.

You will learn about each of these tools as you complete the tutorials in this manual.

Menu Commands

The tool bar is one kind of menu. Most applications provide their commands on menus that appear on the menu bar along the top of the application window.

The principal type of menu is called a *pull-down menu.* To pull down a menu, first point to its title in the menu bar at the top of the screen. **W** Click the mouse button to open the menu. Click on the menu heading, or elsewhere on the screen, to close the menu without choosing an item. **M** Click and hold the mouse button to open the menu. The menu closes when you release the mouse button.

Choosing commands from menus

When a menu is open, you can choose an item from it. Menu items are most often commands. However, they can also be a list of open windows or files, or the names of other menus (*cascading menus*) that list additional commands. You choose a command with the menu pulled down. To choose a menu command, first pull down the menu by pointing to its title and holding down the mouse button. Drag the pointer to the command you want. As you drag, each command that is not dimmed becomes highlighted when the pointer is on it. (Dimmed commands are not available and thus cannot be chosen.) Release the mouse button to choose the highlighted command. The command blinks in the menu, then closes, and the command takes effect.

> **W** *Tip* To move directly to a command on a menu, type the letter or number that is underlined in the item name.

Using a submenu

You also choose commands from submenus by dragging, but the procedure is a little different.

To use a submenu, first pull down its corresponding menu. A small triangle at the right of a menu item shows that that item has a submenu. Drag the pointer to the item with a submenu you want to open. When the item is highlighted, the submenu opens. Drag sideways onto the submenu and then down until the command you want becomes highlighted. Release the mouse button to choose the highlighted submenu command.

Other menu conventions

An ellipsis (...) to the right of a command

A dialog box appears when you choose a command with an ellipsis. A dialog box contains options you need to select before the command can be carried out.

A ✓ to the left of a command

A ✓ indicates that the command is a toggle switch. The check mark means that it is in effect (i.e., on). When you remove the check mark (by choosing the command again), the command is no longer in effect.

A key combination to the right of a command

The key combination is a shortcut for choosing the command. You can press the keys listed to choose the command without first opening the menu.

Working Model Menus

Working Model provides a standard *menu bar* with pull-down menus. The menu bar lists the available menus from which you can choose commands. The Working Model menus are explained in detail below, in the order in which they appear, from left to right.

W The Control menu

The Control menu box is in the upper-left corner of each window. As shown in Figure GS2.7, you use the Control menu to move, size, and close a window.

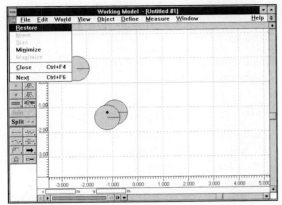

Figure GS2.7

The File Menu

The File menu allows you to open, close, save, print, import, and export a file, and exit from Working Model. The Windows version of this menu is shown in Figure GS2.8. Its options are described below.

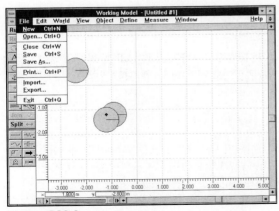

Figure GS2.8

New

This menu option creates a new, blank Working Model document that uses the current default settings.

Open...

This menu option opens a previously created Working Model document. You can have multiple documents open at once.

Close

This menu option closes the currently active document. If there are changes that need to be saved, you are prompted to do so.

Save

This menu option saves the active document to disk. Previously saved documents are updated.

Save As...

This menu option lets you save a copy of an existing document under a new name.

Print...

This menu option displays the Print dialog box, allowing you to print your simulations.

M Page Setup...

This menu option displays a dialog box that specifies printing options, such as paper size and orientation. The contents of the dialog box are dependent on the currently chosen printer.

M Show Page Breaks

This menu option displays page break lines in the workspace. Use page breaks to determine how a document will print.

Import...

This menu option displays a dialog box for importing external files into the workspace. Working Model can import data in the following formats: DXF and 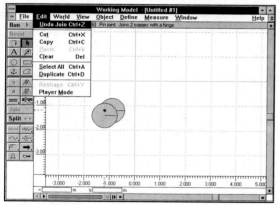 Lincages.

Export...

This menu option displays a dialog box for exporting Working Model data. Working Model can export data in various formats.

W Exit or M Quit

This menu option exits Working Model.

The Edit Menu

The Edit menu allows you to perform basic editing functions on your simulation, and to switch between Player mode and Edit mode. This menu is shown in Figure GS2.9.

Figure GS2.9

Undo

This menu option reverses the last action performed in the simulation. The menu item shows the last action taken, or shows Can't undo if the action is irreversible.

Cut

This menu option removes the selected objects from the document and places them on the Clipboard.

Copy

This menu option places a copy of the selected objects on the Clipboard.

Paste

This menu option pastes the contents from the Clipboard into the active simulation.

Clear

This menu option deletes the selected objects from the simulation without placing the contents onto the Clipboard.

Select All

This menu option selects all of the objects in the active simulation window.

Duplicate

This menu option creates a copy of the selected object or objects.

Reshape

This menu option allows click-and-drag editing of polygons and curved slots to change their shapes.

Player Mode

This menu option is a toggle command that reduces or expands the menu structure of the Working Model program. Simulations saved as Player Documents are perfect for users who will not be editing the simulation. In Player mode, the tool bar on the left of the screen is hidden, making more space for the document on the screen. All the commands you need to use while running a simulation appear on a reduced menu set. When Player mode is active, Edit Mode appears on the menu.

The World Menu

The World menu allows you to control the forces that act on your simulation. Changes made using the commands on this menu are global and apply to all elements of the simulation. This menu appears in Figure GS2.10.

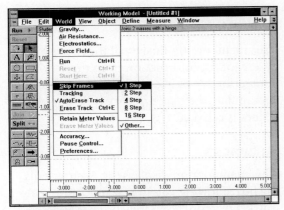

Figure GS2.10

Gravity...

This menu option displays the Gravity dialog box, allowing you to select and control various types of gravity within the active simulation.

Air Resistance...

This menu option displays the Air Resistance dialog box, allowing you to change the air resistance within the active simulation.

Electrostatics...

This menu option displays the Electrostatics dialog box, allowing you to turn on and change electrostatic forces within the active simulation.

Force Field...

This menu option displays the Force Field dialog box, allowing you to create your own custom force fields that will act upon all of the mass objects within the active simulation.

Run

This menu option starts a simulation. When a simulation is in progress, the Run option is replaced by a Stop option. Selecting the Stop option or clicking in the workspace stops the simulation. You can resume the action of the simulation from its last frame (i.e., where you stopped it) by selecting Run again.

Reset

This menu option stops your simulation (if it has not already been stopped) and returns it to its initial conditions and first frame.

Start Here

This menu option sets the starting point of your simulation to the current conditions. A new set of initial conditions is created, based upon the current position and velocity of all objects.

Note: You cannot undo Start Here, which erases your simulation history, including the initial conditions that you specified previously.

Skip Frames

This menu option lets you specify various playback rates for your simulations. Skipping more frames gives you a faster playback on a previously calculated simulation. A submenu appears that lists 1 Step, 2 Step, 4 Step, 8 Step, 16 Step, and Other... The Other dialog box lets you choose your own skip rate. A skip rate of 1 plays every frame of your simulation.

Tracking

This menu option leaves a trace of your simulation at various time intervals. A submenu appears that lists Off, Every frame, Every 2 frames, Every 4 frames, Every 8 frames, Every 16 frames, Every 32 frames, and Other..., as shown in Figure GS2.11. The Other dialog box lets you choose your own custom tracking rate.

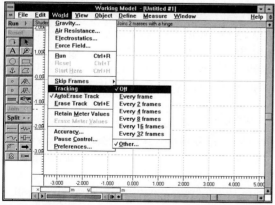

Figure GS2.II

AutoErase Track

This menu option is a toggle switch. If active, it erases the tracks every time you click Reset on the tool bar. A check mark appears next to the option when it is active.

Erase Track

This menu option erases the trace of any tracked simulation.

Retain Meter Values

This menu option is a toggle switch. If active, it keeps the meter data obtained from multiple simulation runs, allowing you to compare the results of more than one simulation on a single meter. A check mark appears next to the option when it is active.

Erase Meter Values

This menu option flushes all the meter history, except the information from the very last simulation. Working Model starts over to accumulate the meter history if Retain Meter Values is active.

Accuracy...

This menu option displays the Accuracy dialog box, allowing you to run simulations more quickly or more accurately by changing the simulation's underlying method of integration and animation.

Pause Control...

This menu option displays the Pause Control dialog box, allowing you to create conditions for your simulations to loop, reset, and pause.

Preferences...

This menu option displays the Preferences dialog box, allowing you to change various program settings to suit the way you use Working Model.

The View Menu

The View menu allows you to control what your simulation looks like in the workspace. This menu is shown in Figure GS2.12.

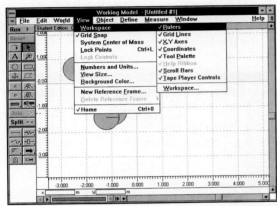

Figure GS2.12

Workspace

The Workspace menu provides control over the appearance of various aspects of the Working Model workspace. Its submenu toggles these options on or off.

Rulers

This menu option toggles rulers on and off in the active simulation window.

Grid Lines

This menu option toggles grid lines on and off in the active simulation window.

X,Y Axes

This menu option shows or hides x, y axes in your simulation.

Coordinates

This menu option shows or hides the coordinate display.

Tool Palette

This menu option shows or hides the tool bar along the left side of the simulation window.

Help Ribbon

This menu option shows or hides the Help Ribbon across the top of the simulation window.

Scroll Bars

This menu option shows or hides the scroll bars.

Tape Player Controls

This menu option shows or hides the Tape Player controls on the bottom of the window.

Workspace...

This menu option displays the Workspace dialog box, which allows you to alter all of the preceding parameters at the same time.

Grid Snap

This menu option is a toggle switch that makes your objects automatically snap to predefined grid lines. A check mark next to the command indicates that Grid Snap is on.

System Center of Mass

This menu option is a toggle switch that shows the center of mass of all mass objects as an X in the simulation window. A check mark next to the command shows the system center of mass.

Lock Points

This menu option prevents points from being moved on mass objects during editing.

Lock Controls

This menu option locks all control objects (buttons, sliders, and meters) to the background. Selecting, resizing, and moving controls is prevented.

Numbers and Units...

This menu option displays the Numbers & Units dialog box, in which you can set a simulation's system of measurement. Unit formats include SI/Metric, English, Astronomical, and CGS.

View Size...

This menu option displays the View Size dialog box, allowing you to set the simulation view and scale to any value.

Background Color...

This menu option selects the color of the background in the Working Model window.

New Reference Frame...

This menu option attaches a reference frame to any selected mass object. You can view simulations from any previously created reference frame.

Delete Reference Frame

This menu option displays a submenu that includes all reference frames except Home. Selecting a submenu item deletes that frame of reference.

List of reference frames

The title of each defined reference frame is appended to the View menu; you can select any reference frame by choosing its title or by using the keyboard shortcut. Note that the Home reference frame is predefined and is always on the list of reference frames.

The Object Menu

The Object menu allows you to control the actions of the objects in your workspace. This menu appears in Figure GS2.13. Menu options not available to the object(s) selected in the workspace are grayed out.

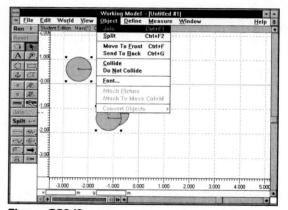

Figure GS2.13

Join

This menu option behaves just as the tool bar command does. It combines two selected elements (points and slots) to form a joint.

Split

This menu option behaves just as the tool bar command does. It separates a joint of other constraints into its elements.

Move To Front

This menu option puts the selected object(s) in front of all other objects.

Send To Back

This menu option puts the selected object(s) behind all other objects.

Collide

This menu option makes all selected objects collide with one another during a simulation.

Do Not Collide

This menu option prevents all selected objects from colliding with one another during a simulation.

W Font...

This menu option displays the Font dialog box, which allows you to select the font type for selected object(s). The fonts, styles, and sizes presented include the fonts installed in your system.

M Font

This menu option chooses the font in which selected text is displayed.

M Size

This menu option chooses the size in which selected text is displayed.

M Style

This menu option chooses the style in which selected text is displayed.

Attach Picture

This menu option attaches a picture to a selected mass object.

Attach to Mass/Detach from Mass

These menu options appear alternately, depending on what objects are selected. Attach to Mass attaches a set of points (including slots, which are actually represented by a key point) to a mass object. Detach from Mass detaches a set of points and slots from the mass object that they are currently attached to. These points and slots are immediately attached to the background.

Convert Objects

This menu option has three options in its submenu. It is grayed out unless a polygon or line segments are selected.

Convert to Lines

This menu option takes selected polygons and converts them to line segments.

Convert to Polygon

This menu option takes selected line segments and converts them to a polygon.

Convert to Curved Slot

This menu option takes selected line segments and converts them to a curved slot. The endpoints of the line segments are converted to the control points of the curve.

The Define Menu

The Define menu is used to created vectors, menu buttons, and controls. This menu appears in Figure GS2.14.

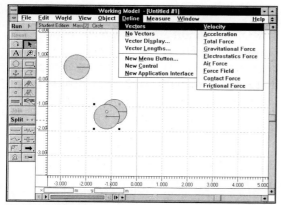

Figure GS2.14

Vectors

The Vectors menu selection determines which vectors are drawn for a selected mass object. You can select any combination of the listed vectors, and the corresponding vectors will be drawn and dynamically updated during the course of a Working Model simulation. For example, the submenu shown in Figure GS2.14 displays the vectors available for a mass object.

No Vectors

This menu option prevents any vectors from being drawn for a selected mass object.

Vector Display...

This menu option displays the Vector Display submenu, in which you can change vector colors and styles.

Vector Lengths...

This menu option displays the Vector Lengths submenu, in which you can globally change the length of all velocity, force, and acceleration vectors.

New Menu Button...

This menu option displays the New Menu Button dialog box, in which you can create buttons that perform menu commands.

New Control

This menu option allows you to add a control to the workspace. Controls allow you to adjust simulation parameters before *and* while a simulation is running. The object properties that can be adjusted by the control appear in a menu and will vary, depending on what kind of object is selected. By default, New Control creates a slider bar control as a data entry tool. Working Model allows more versatile input tools, such as text, button, and external file input. The new control options for a mass object appear in Figure GS2.15.

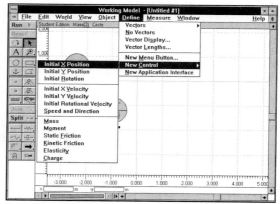

Figure GS2.15

New Application Interface

This menu option allows you to have Working Model link its data output to other application programs. Two examples of applications are MATLAB (Windows only) and Excel. See Real Time Links with External Applications in Appendix C for more information.

The Measure Menu

The Measure menu allows you to create meters to measure properties of a Working Model simulation. This menu appears in Figure GS2.16. Because a circle is selected in the workspace, the items showing are those for a mass object.

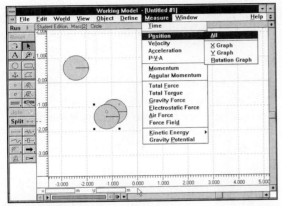

Figure GS2.16

Time

Time creates a meter that measures the time elapsed in the simulation. Time is the only meter that is always available.

List of other measurements

The Measure menu lists the valid measurements for any selected object.

The current display in Figure GS2.16 shows valid measurements for a single mass object. These include Position, Velocity, Acceleration, PVA (Position, Velocity, and Acceleration in one meter), Momentum, Angular Momentum, Total Force, Total Torque, Gravity Force, Electrostatic Force, Air Force, Force Field, Kinetic Energy, and Gravity

Potential. Selecting other objects gives you various measurement options.

When you select two mass objects, the Measure menu changes so that you can measure forces that inherently act between a pair of objects, including Contact Force, Friction Force, and Pair Gravity Force.

- Click on one circle in your workspace.
- Select the **Measure** menu to observe the meter options available to you for one mass object.
- Close the **Measure** menu.
- Holding down the **Shift** key, click on another circle in your workspace.
- Select the **Measure** menu to observe the meter options available to you for a pair of mass objects.
- Close the **Measure** menu.
- Click on an empty spot in the workspace to deselect all objects.
- Select the **Measure** menu to observe that the only meter option available to you is Time.
- Close the **Measure** menu.
- Click on the pin joint in your workspace.
- Select the **Measure** menu to observe the meter options available to you for this constraint.
- Close the **Measure** menu.
- With the pin joint still selected, click on **Split**.
- Click on an empty spot in the workspace to deselect the joint.
- Click on the point on the top circle in your workspace.

- Select the **Measure** menu to observe the meter options available to you for a point.
- Close the **Measure** menu.

The Window Menu

The Window menu provides three utility windows to specify precise values for an object's properties. This menu appears in Figure GS2.17.

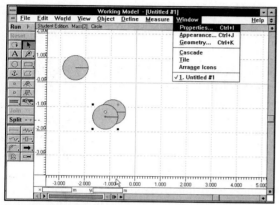

Figure GS2.17

Properties...

The Properties... utility window provides direct access to the physical properties of the currently selected object. Different fields appear, depending on the type of object selected. You can change the properties of several selected objects at the same time: select multiple objects and modify the desired properties in the window.

Appearance...

The Appearance... utility window controls the appearance of selected objects. Color, fill,

tracking, and center of mass display are controlled by this window.

Geometry...

The Geometry... utility window controls the geometry of selected mass objects. The properties that appear in this window depend on the type of object selected. A rectangle's geometry is specified by width and height. A polygon's vertices can be altered by editing the values in this window.

ⓦ Cascade

This menu option arranges all the currently open document windows in a cascaded fashion (the title bar for each window is visible).

ⓦ Tile

This menu option arranges all the currently open document windows in a tiled fashion (each window is visible but reduced in size).

ⓦ Arrange Icons

This menu option neatly aligns the iconized Working Model documents.

List of open documents

A list of open documents is appended to the bottom of the Window menu. All current documents are listed in alphabetical order. They are numbered 1 through n (with a maximum of 10 documents) and you can access them by selecting the desired number. The active window is identified by a check mark next to its name. ⓜ On the Macintosh, this list appears only if you have more than one document open.

W The Help Menu in Windows

The Help menu allows you to access Working Model's on-line help facility. This menu appears in Figure GS2.18.

M Help is not available on the Macintosh.

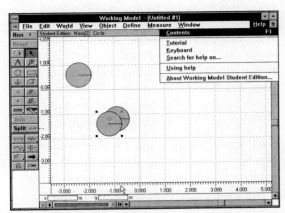

Figure GS2.18

Contents

This menu option displays a list of the main help topics. You can obtain more detailed information by traveling down the help structure.

Tutorial

This menu option puts you into the Working Model hands-on tutorial.

Keyboard

This menu option gives you information related to keyboard usage and shortcuts.

Search for help on...

This menu option allows you to look for help on a specific topic. A list of topics is presented alphabetically, and for each topic a list of help screens that discuss the topic is displayed.

Using help

This menu option provides information on the help system.

About Working Model Student Edition...

This option displays information about Working Model, including its version, copyright, licensee information, and system statistics.

Working with Dialog Boxes

Working Model displays a dialog box when it needs additional information to complete a task. An ellipsis (...) after a menu command indicates that a dialog box will appear when you choose that command.

Most dialog boxes contain options you can select. After you specify options, you can choose a command button to carry out the command.

- Select **View** menu, **Numbers and Units...**

The Numbers & Units dialog box should appear on your screen, as shown in Figure GS2.19. Keep this dialog box open so that you can refer to its components and experiment with it as you learn more about Working Model dialog boxes.

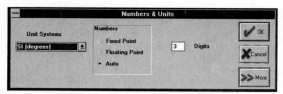

Figure GS2.19

If a dialog box has a title bar, you can move the box to another location on your desktop. Moving a dialog box is just like moving a window: you drag the title bar to move the dialog box to its new location.

Moving within a dialog box

Often you need to move within a dialog box to select one or more options. The currently selected option is marked by the selection cursor, which appears as a dotted rectangle, a highlight, or both.

To move within a dialog box, click on the option or area you want to move to, or **w** press Tab to move clockwise or Shift+Tab to move counterclockwise through the options or areas.

> **w** *Tip:* If the option, box, or button has an underlined letter in its name, you can choose that item by pressing and holding down ALT while typing the underlined letter to get to that option or area.

Types of options

Several types of dialog box options appear in Working Model and are described in the following sections.

Pull-down menus

A *pull-down menu* appears initially as a rectangular box containing the current selection. In Figure GS2.19, under the heading **w** Unit

Systems or **M** Unit System, the current selection is SI (degrees). When you select the down arrow in the square box at the right, a list of available choices appears. If there are more items than can fit in the box, scroll bars are provided. Usually you can select only one item on a pull-down menu.

Command buttons

You choose a *command button* to initiate an action, such as carrying out or canceling a command. The OK and Cancel buttons are common command buttons. They are often located along the bottom or on the right side of the dialog box, as they are in Figure GS2.19.

Some dialog boxes contain a Help command button and/or a **w** More or **M** More choices command button. This button expands the active dialog box. Figure GS2.20 shows the expanded Numbers & Units dialog box after the **w** **More** or **M** **More choices** command button was selected.

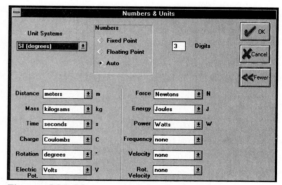

Figure GS2.20

The currently selected button has a darker border than other buttons. You can choose the selected button simply by pressing Enter.

- Click **Cancel** to close the Numbers & Units dialog box.

- Select **World** menu, **Accuracy...**
- Select the **W** More or **M** More choices command button.

The expanded Simulation dialog box should appear on your screen, as shown in Figure GS2.21.

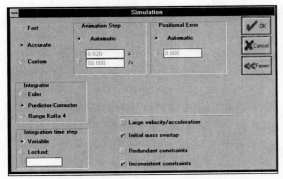

Figure GS2.21

Option buttons

Option buttons represent a group of mutually exclusive options. You can select only one option at a time. If you already have one option selected, your current selection replaces it. The selected option button contains a black dot. The Fast, Accurate, and Custom choices in Figure GS2.21 are option buttons.

Text boxes

You type information into a *text box*. When you move to an empty text box, an *insertion point* (a flashing vertical bar) appears. The text you type starts at the insertion point.

If the box you move to contains text, and the text is highlighted, any text you type will replace it. You can also delete the existing text by pressing DEL or Backspace.

- Select the button under the Automatic button in the Animation Step field.
- Experiment with entering type into these type boxes.

Check boxes

A *check box* means you can select or clear an option. You can select as many check box options as needed. When a check box is selected, it contains a **W** ✓/**M** ×. In the lower portion of the Simulation dialog box shown in Figure GS2.21 are check boxes for Large velocity/acceleration, Initial mass overlap, Redundant constraints, and Inconsistent constraints.

> *Tip:* If the option name has an underlined letter, you can select or clear the check box by pressing and holding down the ALT key while typing the underlined letter.

Closing a dialog box

When you choose the OK command button, the dialog box closes and the command is carried out. You can also select Cancel to close the dialog box and cancel the command, as you did when closing the Numbers & Units dialog box.

- Click the **Cancel** command button to close the Simulation dialog box.

Using an alert box

Some dialog boxes may display information, warnings, or messages indicating why a requested task cannot be accomplished. One of these, the *alert box*, is designed to alert you that you may accidentally lose some of your work, ignore constraints while running an experiment, or quit your Working Model application program without saving your simulation. Its message warns you that you may be doing something you didn't intend and asks you to confirm that you want to go ahead with the action. On some systems, alert boxes are accompanied by a sound to warn you of a possible mistake.

You can press Return or Enter, or click OK to continue, or Cancel if you made an error.

Keyboard Functions

You can accomplish most tasks by using either the mouse or the keyboard. Using a mouse is usually easier and faster, however. Many commands have a keyboard equivalent, usually made up of the **w** Control or **M** Command key and one or more other keys. Each of the *keyboard shortcuts* shown in Table GS2.1 is displayed to the right of the command it represents on the menu.

Using modifier keys

The following list explains modifier keys that you can use when editing objects.

The Shift key

You can use the Shift key to select more than one item by holding down the Shift key while clicking on the items you want. Clicking on an already selected object while holding down the Shift key deselects the object.

The Tab key

You can use the Tab key to navigate from one text field to the next in a utility window. You can also use the Tab key to select a utility window. If you have an object in the workspace selected, pressing the Tab key automatically

selects the last-selected utility window associated with the object.

w The Control key

w Holding down the Control key while dragging the endpoint of a constraint will maintain its current length. Control-drag will also maintain the current connections of constraints and mass objects.

M The Option key

M To maintain the rest length of a constraint while resizing, hold down the Option key.

M The Command key

M To maintain the current connections of constraints and mass objects when dragging, hold down the Command key.

The Smart Editor

The Smart Editor™ is the core of the Working Model user interface that allows you to use the mouse to manipulate your model or screen without disturbing connections and constraints between objects. Although you create objects with the same kind of click-and-drag motion used by drawing programs, the Smart Editor maintains the fundamental integrity of the components and of the joints between them. You can position objects by using the standard click-and-drag paradigm, or by specifying their coordinates precisely in utility windows. In all cases, the Smart Editor makes sure that no link is broken and no mass is stretched.

Here's how the Smart Editor works: when you drag a mechanism with the mouse, it moves like a real mechanism.

Table GS2.1 Keyboard Shortcuts

Key	Action
a, A	Selects the Anchor tool
r, R	Selects the Rotate tool
Space bar	Selects the Arrow tool
Z	Selects the Zoom out tool
z	Selects the Zoom in tool
+	Steps forward one frame
-	Steps backward one frame
(W) Control-C or Control-Insert (M) F3 or Command-C	Copy
(W) Control-X or Shift-Delete (M) F2 or Command-X	Cut
(W) Control-F1 (M) Command-+	Join
(W) Control-V or Shift-Insert (M) F4 or Command-V	Paste
(W) Shift-Control-R (M) Shift-Command-R	Runs from the last computed frame
(W) Control-F2 (M) Command--	Split
(W) Control-Z or Alt-Backspace (M) F1 or Command-Z	Undo
(W) Delete	(W) Clear
(W) F1	(W) Help
(W) Alt-Enter, Ctrl-I	(W) Invokes the object's Properties window
(W) F2	(W) New document
(W) Control-F12	(W) Open
(W) Control-Shift-F12	(W) Print
(W) Alt-F4	(W) Quit
(W) F5	(W) Run/Stop
(W) Shift-F12	(W) Save
(W) F12	(W) Save As

- Click and drag on one of the circles connected with the pin joint.

It is difficult to see any behavior that can be attributed to the Smart Editor because both circles move in their present position to wherever you move the cursor.

- Select the **Anchor** icon from the tool bar.

- Click on one of the circles to anchor it to the background.

- Click and drag on the circle *without the anchor* to see how it behaves.

The circle should be able to rotate to new positions, never losing contact with its joint, and never causing the anchored circle to move.

The Smart Editor enforces constraints while you edit. It prevents a mechanism from disintegrating when its components are moved around. Instead, other components are moved or rotated (subject to their own constraints) until the desired move is accomplished. Sometimes a drag or rotation may be inconsistent with the constraints that are imposed, in which case you will get a warning message on your screen that constraints are being ignored or that Working Model cannot perform the action.

Clicking and dragging

The Smart Editor is designed to follow the click-and-drag paradigm as much as possible. If possible, the mechanism directly follows the pointer. The point at which you click follows the current pointer position.

If constraints make this impossible, then the mechanism is moved so as to make the initial click point as close as possible to the current pointer position.

The rules that the Smart Editor uses in moving objects in the workspace are simple and consistent. The easiest way to understand the Smart Editor is to play with it. The Smart Editor is transparent, and won't be obvious because it is intuitive and consistent with everyday experience. However, understanding its rules of behavior will help you as you complete the tutorials. You may want to refer to this section again after you have completed a tutorial or two.

Rule # 1: No constraint is broken during editing. If you drag a rectangle that is joined to a circle, the circle must follow along.

Rule #2: Endpoints of constraints cannot move on the objects they are attached to during editing. Points that define a joint do not move relative to the object they connect to. Joints must remain in place.

Rule #3: If you simultaneously select a collection of objects, a drag or rotate operation will treat them as a rigid unit, so that no alteration in their relative positions or rotations will occur. This behavior is consistent with what one would expect.

Rule #4: Collisions are ignored during editing.

Rule #5: No joint will rotate unless some constraint forces it to do so during editing.

When is the Smart Editor used?

The Smart Editor is automatically used to resolve conflicts between user commands and constraints in the following situations:

- When an object is dragged or rotated, the Smart Editor dynamically updates the workspace. A moving picture of your mechanism follows the pointer as you drag or rotate.

- When the Join command is invoked, the Smart Editor verifies that no constraint is being violated. Because the Join command can cause mass objects to move around in the workspace, the Smart Editor moves objects as necessary to satisfy constraints.

- When a new coordinate position is entered into the Properties utility window, the Smart Editor verifies that the new position is not inconsistent with pre-existing constraints.

Select-all-drag

There is one major exception to all of these rules—the select-all-drag exception. If every constraint attached to a selected mass object is also selected, dragging the mass object disables the Smart Editor.

Joining and splitting

The Smart Editor can automatically assemble or disassemble a mechanism. You can temporarily split pin joints, leaving a separate point on each mass object. You can edit these points individually, and then reassemble the pin joint with the Join command.

Controlling the movement of objects

When you are editing complex mechanisms with the mouse, or when joining, it is sometimes useful to lock down certain parts of the mechanism temporarily. If no part of a mechanism is joined to the background, dragging any component in the mechanism will move the whole mechanism, and joining will move any object to achieve assembly. You can use the Anchor tool to lock objects in place. This

will ensure that only the unanchored objects are moved by the mouse or by the Join tool.

Using the Smart Editor with ropes

A rope object is a device that introduces non-linear equations into the Smart Editor. When you are dragging mechanisms that contain many ropes, the Smart Editor may lock ropes in their fully extended positions. If an extended rope will not go slack, release the mouse button and then continue dragging.

Precision numerical assembly

The Smart Editor assembles mechanisms based on numerical values. Whenever you enter the position of a mass object, point, or joint, the Smart Editor makes sure that joints are not broken. If necessary, the Smart Editor will move other mass objects to keep the integrity of all joints in a mechanism.

What if the Smart Editor fails?

Situations can arise in which it is impossible to satisfy all of the constraints imposed on a system.

Trying to drag an object to a position inconsistent with the constraints does not cause an error message. The Smart Editor will try to find the best solution by moving objects to minimize the distance between the pointer and the place on the object where the pointer was originally clicked. If you try to drag an object too far, it will follow the mouse, and then stop after going as far as it can. If you try an edit that isn't completed by Working Model, check to see that the movement doesn't violate a constraint of the mechanism.

Methods of Numerical Integration

Working Model calculates the motion of interacting bodies using numerical methods. The simulation engine breaks a simulation problem into discrete time segments, within which Working Model computes motion and forces, while making sure that all constraints are satisfied. Working Model allows you to control the accuracy and speed of the simulation by choosing among different methods of numerical integration and by setting the size of the time step. You will see how this is done in the tutorials in this manual. For complete information on the numerical integration methods used by Working Model, refer to Appendix A.

Exiting Working Model

To exit Working Model:

- Select **File** menu, **W** **Exit** or **M** **Quit**.

A message window appears on your screen, asking if you want to save your file. You need not save this simulation.

- Select **W** **No** or **M** **Don't Save**.

Working Model quits to **W** Microsoft Windows or **M** the Finder.

Tutorials

Projectile Motion

Objectives

In this tutorial, you will learn to:
- Define workspace settings
- Create circles
- Specify an initial velocity
- Zoom
- Run the experiment
- Move the view across the workspace
- Create vector displays
- Create and locate output meters
- Use the Tape Player
- Use air resistance
- Save an experiment
- Exit Working Model

Dynamics principles related to this tutorial

- Position, velocity, and acceleration of particles
- Projectile motion
- Rectangular components of vectors
- Acceleration due to gravity
- Air resistance

Introduction

In this first tutorial, you will use Working Model to describe the motion of a particle in a gravitational field. This type of motion is called *projectile motion,* and can be used for many engineering applications, from predicting how far you can drive a golf ball to calculating how an airplane can accurately target the drop of a relief shipment so that it lands precisely where it is needed. You will also begin to learn some of the basic concepts of developing and running dynamic simulations using Working Model.

Problem Statement

You are designing a sports training machine that will kick soccer balls from the ground into the air toward the goal. Your preliminary design gives each ball an initial velocity of 15 m/s at the angle shown in Figure 1.1. Assume that you can neglect air resistance. What are the height (*h*) and the range (*d*) of the ball? How long does it take to return to the ground?

You will create a Working Model simulation for this situation, showing the horizontal and vertical components of velocity as the ball travels. You will also display the ball's acceleration as it moves.

Before you examine the dynamics of the soccer ball, you will learn some of the basic Working Model tools necessary to create this simulation.

Figure 1.1

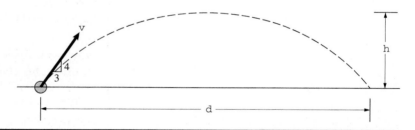

air resistance neglected

$V_1 = 15 \, m/s$

Starting Working Model

The Student Edition of Working Model should be installed on your computer before you begin. If it is not, refer to Chapter 1, Installing Working Model, in Part 1, Getting Started.

Start Working Model by double-clicking on its icon. If you need help, refer to Chapter 2, Working Model Basics, in Part 1, Getting Started.

Working Model prompts you to enter a three-digit code from one of the pages of this manual. These three digits appear along the inside edge of the bottom of the page. At the prompt, locate the appropriate page, type the three digits in the order in which they appear, and click **OK.**

A new file called Untitled#1 opens in the Working Model window. Your screen should look like Figure 1.2.

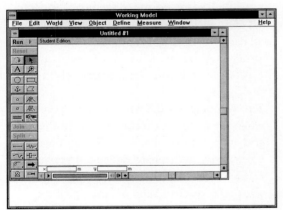

Figure 1.2

Setting Up the Workspace

Working Model provides several tools to make it easier to create and position simulations. These tools are located on the Workspace submenu of the View menu. X,Y Axes, Grid Lines, and Rulers are aids that you will use regularly as you create simulations. *X,Y Axes* are a pair of solid lines that intersect at (0,0), the *origin* of the axes. *Grid Lines* are dotted lines that are spaced at equal intervals on the screen. *Rulers* appear at the left edge and bottom of the workspace.

Defining workspace settings

Turn the drawing axes, grid, and rulers on now.

- Select **View** menu, **Workspace, X,Y Axes.**
- Select **View** menu, **Workspace, Grid Lines.**
- Select **View** menu, **Workspace, Rulers.**

  Note: These instructions assume that you have just installed Working Model as described in the Getting Started portion of this manual, and that you are changing the settings for the workspace for the first time. If you have changed your workspace settings before, or if someone else may have used your Working Model, refer to Figure 1.3, which shows all the settings that should be on (have a check mark to the left of them) under View menu, Workspace: Rulers, Grid Lines, X,Y Axes, Coordinates, Tool Palette, Help Ribbon, Scroll Bars, and Tape Player Controls.

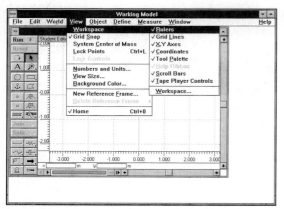

Figure 1.3

The ruler is numbered at every meter. The smaller hash marks represent 0.1-meter intervals. The grid line appears at 1-meter intervals.

You can enlarge your Untitled#1 window so that it fills your entire Working Model window.

W Click on the **Maximize** button (the up arrow) in the top right corner of your Untitled#1 window .

M Click on the **small box** in the top right corner of your Untitled#1 window.

Your screen should look like Figure 1.4.

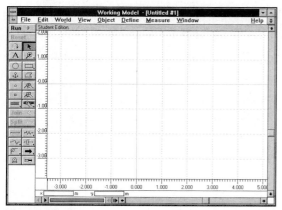

Figure 1.4

Grid snap

Accurate placement of objects in Working Model is difficult without the use of special drawing aids. The Grid Snap feature allows you to accurately align objects to the grid. Grid Snap is located on the View menu; a check mark to its left indicates that it is activated. If it is not already on, turn Grid Snap on now. If a check mark appears next to Grid Snap, choosing it will turn it off. Be sure you turn it on.

- Choose **View** menu, **Grid Snap.**

Grid snap points are located in the same positions as the smallest marks on the rulers. The default Working Model grid snap is set at 0.1-meter intervals. When Grid Snap is on, your mouse cursor snaps to the grid snap points in your workspace. The location of your cursor is indicated by dotted lines on the rulers. Watch the cursor location snap to these increments as you pick points on the screen. The last point picked also appears on the rulers to help you gauge distance between picked points. Pick some points to see how the dotted rules change. If you do not use grid snapping to place objects, you will probably have to use another method (which you will learn later in this tutorial) to locate objects precisely.

> *Tip* Although Grid Snap is a useful tool when trying to pick a coordinate on the grid, it sometimes makes placing an object at a specific coordinate difficult. If you have trouble choosing a specific point, you may want to turn Grid Snap off, then pick the point.

Coordinate display

The coordinate display at the bottom of your screen can also be used to identify the position of your cursor. Pick a point on the screen to see its absolute location displayed in the boxes labeled x and y at the bottom of the document. If you click and drag (with the mouse button held down) the coordinate display changes to show the displacement relative to the initial position. The coordinate display shows (from left to right) the displacement in terms of the x and y values, the total distance from the starting point, and the displacement angle. This is a feature you will use in future tutorials to help you reposition and resize objects.

To save these settings so that they will appear each time you start Working Model, choose Save Current Settings from the Preferences dialog box.

- Select **World** menu, **Preferences...**
- Choose **Save Current Settings.**
- **w** Click **OK.**

  Note: Save Current Settings on the Windows platform does not include the window size. You will need to resize your window as needed each time you open Working Model.

m A warning box appears that asks: Use settings of "Untitled#1" for new documents? (This includes the window size and settings under the View and World menus.)

m Choose **Save.**

m Click **OK.**

Your workspace will open exactly as it is now when you open future sessions of Working Model.

Creating the Components

The components of your Working Model simulation are masses, constraints, and points. *Masses* include circles, rectangles, squares, and polygons. These are used to model bodies and may be referred to as bodies, entities, mass objects, or just objects in these tutorials. *Constraints* include springs, ropes, dampers, actuators, motors, forces, joints, torques, rods, and separators. *Points* are either isolated points or endpoints of a constraint. You will cover masses (a circle) in this tutorial. Other components will be dealt with in detail in future tutorials.

Creating circles

The Circle icon on the tool bar (it looks like a circle) is used to create circles of different sizes in your workspace. If you click once on this icon, it becomes highlighted (turns gray) and allows you to create one circle. If you double-click on it, it turns black; you can then create as many circles as you wish. (If an icon turns black when you click on it, that tool is active until you choose another icon.) If you simply click on the workspace, you get a circle of standard size, preset by Working Model; if you click on and drag the circle at a corner, you can make the circle any size you wish.

- Click on the **Circle** icon on the tool bar.
- Click at the **origin (0,0)** of the drawing axes.

Your workspace should now look like Figure 1.5.

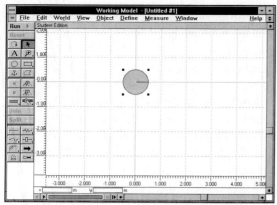

Figure 1.5

Specifying an Initial Velocity

Vectors represent the properties of velocity, acceleration, and force as visual arrows. The direction of the arrow shows the direction of the velocity, acceleration, or force. The length of the arrow corresponds to the magnitude of the velocity, acceleration, or force. You will set up a vector to specify the initial velocity of the circle at 15 m/s.

Working Model lets you create a velocity vector in several ways. One way to create an initial velocity vector is to click on the center of the object and drag in the direction that you want the object to move. You will learn another method of creating vectors later in this tutorial. Use Figure 1.6 as a guide when you position your initial velocity vector.

- Click on the circle you just created to select it, if necessary. (When an object is selected, its handles appear as four square dots.)

- Click on the dot indicating the center of the circle and drag it to a point above and to the right of the circle; release the mouse button.

An arrow indicating the initial velocity of the circle appears, as shown in Figure 1.6.

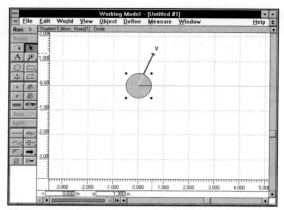

Figure 1.6

The Properties Window

Double-clicking on an entity opens its Properties window; you can also open this window by first selecting the desired object, then selecting the Window menu, and then selecting Properties...

- Double-click anywhere on the circle.

The *Properties window* contains, in list form, various pieces of information about the selected object. Starting at the top of the window, the particular body, in this case Mass[1], is identified. Next, it is characterized as a circle. The next three items locate the body's mass center and angle of rotation (measured counterclockwise from the positive x axis). If you placed your circle at the origin of the x,y axes with Grid Snap on, your x and y coordinates for the body's mass center should be 0 m and 0 m. Your angle of rotation (Ø) should be 0°. Next, the x and y components of the initial velocity of the mass are shown, followed by the initial angular velocity. Additional material and geometric properties, which will be treated in later tutorials, are also listed. The Properties window is context sensitive; each object's Properties window displays different information. You will also use Properties windows with other types of objects.

Identifiers

Working Model automatically assigns identifiers to each object created. There are five types of identifiers: Mass[], Point[], Constraint[], Output[], and Input[]. Every identifier also contains a number within its brackets. The numbers distinguish among each of the masses, points, constraints, outputs, and/or inputs, and are assigned according to the order in which objects are created.

Mass[1]   Mass[] identifies mass objects, such as circles, polygons, and rectangles. Mass[1] is the ID for mass #1.

Point[11]   Point[] identifies point objects. Point objects are either isolated points or the endpoints of a constraint. Point[11] is the ID for point #11.

Constraint[44]   Constraint[] is the identifier for constraint objects, including springs, ropes, joints, and pulleys. Constraint[44] is the ID for constraint #44.

Output[12]   Output[] is the identifier for all meters. Output[12] is the ID for meter #12.

Input[5]   Input[] is the identifier for all input controls, including sliders, text boxes, and buttons. Input[5] is the ID for input control #5.

- Using the Properties window, make sure that the values of **x** and **y**, the coordinates of the mass center, are **(0,0);** if necessary, highlight the text box in the boxes next to x and y and change each to **0.**

- Highlight the text box next to V_x.

- Change the value to **9** m/s.

- Highlight the text box next to V_y.

- Change the value to **12** m/s.

- Click on another text box or press **Enter.**

These values, 9 m/s and 12 m/s, are the x and y components of the 15 m/s initial velocity given for this problem. Your workspace should now look like Figure 1.7.

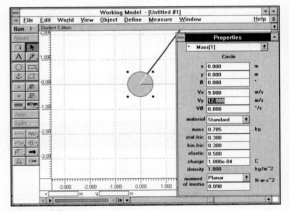

Figure 1.7

Note: Your initial velocity vector will only be visible when the object is selected.

W Double-click in the upper-left-hand corner of the Properties window to close it.

M Click in the upper-left-hand corner of the Properties window to close it. (You must select the window before you can close it.)

Note: In the remainder of the tutorials, you will simply be instructed to close the window; the different steps for Windows and Macintosh will not be spelled out.

Tip Many Working Model menu items can be chosen with keystrokes instead of the mouse. These are called *keyboard shortcuts*. The keyboard shortcut for each menu item is listed to the right of the selection on the menu. A complete list of keyboard shortcuts is listed in the Getting Started section of this manual.

The keyboard shortcut for opening the Properties window is **W** the Control key and the letter I (Ctrl+I) or **M** the Command key and the letter I (Command+I). Remember that the object must be selected first to indicate *which* object's properties are to be displayed.

Zooming

The Zoom icon can enlarge or shrink the appearance of the objects in the workspace. Zoom doesn't change the actual size of the object. The Zoom icon is located on the tool bar; it looks like a magnifying glass. The Zoom icon has a small arrow on the right side of it; this indicates that it has more selections than you see now. If you hold down the mouse button on the Zoom icon, you will see two options: Zoom In, a magnifying glass

with a plus in it; and Zoom Out, a magnifying glass with a minus in it. When you activate the Zoom In icon and move the cursor to the workspace, the cursor will look like a magnifying glass with a plus in it; each time you click on it, it magnifies all objects in the workspace. When you select the Zoom Out icon and click in the workspace, all objects in the workspace appear smaller. Once either Zoom icon has been selected, you can hold the Shift key down to toggle to the other Zoom option.

The icon turns black when you activate zooming and it remains active until you choose another icon.

- Click on the **Zoom** icon on the tool bar.

- Hold the **Shift** key down to activate the **Zoom Out** option.

- Click twice on the workspace (not the circle).

- Click on the **Arrow** icon on the tool bar to deactivate Zoom.

Figure 1.8 shows the workspace after the cursor has been clicked twice.

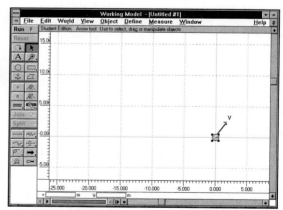

Figure 1.8

Running the Simulation

You can run a simulation at any time to see how the masses will behave. To run a simulation, click on Run on the tool bar, or choose Run from the World menu. While the simulation is running, the Run button becomes a Stop button and pauses your simulation if selected. If you choose Run again, the simulation continues to run from the point at which you stopped it. Using the Reset button on the tool bar or choosing Reset from the World menu resets your simulation to its initial conditions, and you can run the simulation again from the beginning.

- Click on the **Run** button on the tool bar.

The circle should begin to move on your screen. The cursor changes to a stop sign while the simulation is running. Clicking the mouse in the workspace pauses the simulation and is equivalent to clicking the Stop button on the tool bar.

- Click on the **Stop** button on the tool bar.

The circle stops its motion and remains where it was when you selected Stop.

- Click on the **Run** button on the tool bar.

The circle continues from where you stopped its motion. Note that the mass disappears from the workspace before the simulation is complete.

- Click on the **Reset** button on the tool bar.

You can select Reset while the simulation is still running to stop the simulation and return the body to its original position and initial conditions. You can also select it after you have

stopped the simulation to return the body to its original position and initial conditions.

Moving the view across the workspace

In order to position the mass so that the entire path of motion is visible when the simulation is run, you will move the view of the workspace.

- Click on the horizontal and vertical scroll arrows at the bottom-right and upper-right edges of the screen until the mass is in the lower left corner.

The workspace should now look like Figure 1.9.

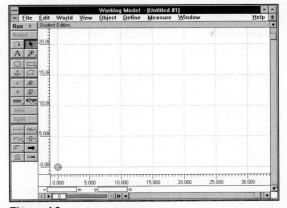

Figure 1.9

- **Run** the simulation again.
- Click on the **Reset** button when the object crosses the x axis again.

Please note that your simulations in Working Model do not end. They continue to run as long as you allow them to. For the purposes of this tutorial, your simulation is "finished" when the circle crosses the x axis again.

Obtaining Results

Working Model allows you to measure many physical properties, such as velocity, acceleration, and energy, by using *meters* and vectors. Meters and vectors provide visual representations of quantities you want to measure.

Creating Vector Displays

Earlier in this tutorial, you specified the *initial* velocity with a vector. Now you will measure results within your simulation by creating vectors to measure the velocity in the x and y directions *as the body moves*. The Define menu is used to create vectors, menu buttons, and controls. The Define menu is context sensitive and the options you are able to select depend on which objects in your workspace are selected. You will use the Define menu to create velocity vectors and to change their display. You will learn more about the other options on the Define menu in later tutorials.

- Click on the circle to select it.

- Select **Define** menu, **Vectors, Velocity.**

This will cause velocity vectors to be displayed on the mass at all times, not only when the mass is selected. Note that the mass has not changed in appearance. The Working Model default setting for Vector Display is to show the magnitude, or total velocity vector, which is the vector you had already created. You will change Vector Display so that your velocity vectors in the x and y directions will show.

- Select **Define** menu, **Vector Display...**

The Vector Display dialog box opens. Vectors can be displayed with their x, y, and total components. Velocity, acceleration, and force vectors can be displayed in different colors. Force vectors can be displayed at their point of application, or at the center of the mass of the object they act upon. For this tutorial, you will display the x and y components of the vectors. You will leave the color and the force vector options set to the Working Model defaults. You will make changes so your screen looks like Figure 1.10.

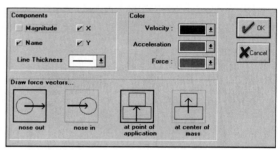

Figure 1.10

- Click on **X** and **Y** in the Components section in the upper-left corner of the dialog box to turn them **on.** (A check mark appears next to them.)

- Click on **Magnitude** so that the total velocity vector is not displayed. (The check mark next to Magnitude disappears.)

- Click **OK.**

The velocity vectors for the x and y directions should now be shown separately, labeled V_x and V_y. Your initial velocity vector has disappeared. Your screen should now look like Figure 1.11.

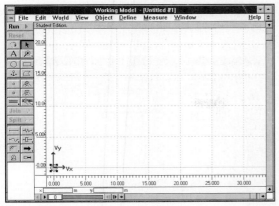

Figure 1.11

- **Run** the simulation again.

As you run the simulation, you will note that the vector for velocity in the x direction remains constant; this is because there is no acceleration in the x direction. The velocity in the y direction changes continuously because the acceleration due to gravity is acting in the y direction.

- Select **Reset** before you continue.

Creating and Locating Output Meters

The Measure menu allows you to create output meters to measure the properties of a Working Model simulation. Measure is context sensitive; the options that appear depend on what objects are selected in your workspace. Output meters can give information in three forms: one numerical and two graphical—line graphs and bar graphs. All of these will be dealt with at greater length in later tutorials. In this exercise you will look at numerical output only.

- Click on the circle in the workspace to select it.

- Select **Measure** menu, **Acceleration, All.**

The output meter that appears shows the digital display. Your screen should now look like Figure 1.12.

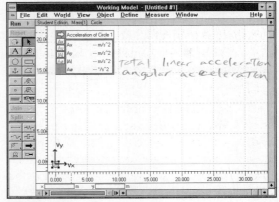

Figure 1.12

In its current form, the Acceleration meter shows the x component of acceleration, the y component of acceleration, the total linear acceleration (|A|), and the angular acceleration of the mass ($A_\emptyset$). The angular acceleration of the mass is assumed to be zero for this problem, since you are treating the mass as a particle. You will deactivate the angular acceleration output quantity because it is of no use in this problem.

- Click on $A_\emptyset$ in the left-most column of the Acceleration meter to deactivate this value. This selection becomes dim and the value disappears from the meter.

Now you will continue to use the Measure menu to create output meters to measure the position of the mass and the time elapsed in the simulation.

- Click on the circle to select it again.
- Select **Measure** menu, **Position, All.**

Sometimes you need to move the meters for easier viewing.

- Click and drag on the title bar of the meter to move it to the top of the workspace and to the right of the Acceleration meter.

When you are finished with this step, your workspace should look like Figure 1.13.

Figure 1.13

Since the rotational motion of the body will always be zero for this problem, you do not need the output meter to display its value.

- Click on **rot** in the left-most column of the Position meter to deactivate the value.
- Select **Measure** menu, **Time.**
- Click and drag on the meter to move it to the top of the workspace and to the right of the Position meter.

> **Tip** Time is the only Measure menu option available when no items are selected in your workspace.

Your workspace should now look like Figure 1.14.

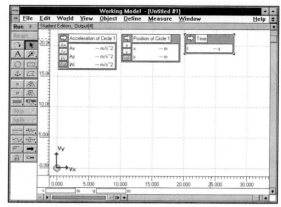

Figure 1.14

- **Run** the simulation again.

Notice the displays in the three output meters as you run the simulation.

- Select **Reset.**

Using the Tape Player

While running a simulation, Working Model also records it, using the *Tape Player* feature. This allows you to play simulations backward, to skip frames of the simulation, and to play simulations more quickly after all calculations have been completed. In addition, you can pause simulations to observe important details. The Tape Player controls also provide a visual indication of the number of frames in the simulation.

The Tape Player controls should appear to the left of the bottom scroll bar, as shown in Figure 1.15; if they do not,

- Select **View** menu, **Workspace, Tape Player Controls.**

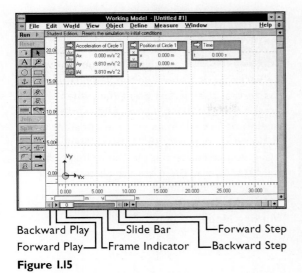

Backward Play ── Slide Bar ── Forward Step
Forward Play ── Frame Indicator ── Backward Step

Figure 1.15

Now the Tape Player controls are at the bottom left-hand side of your screen. Starting from the left, there is the *Backward Play* button, the *Forward Play* button, the *Frame Indicator and Slide Bar*, the *Backward Step* button, and the *Forward Step* button. With the Tape Player, you can:

1. Run the simulation backward and forward by clicking on the arrows to the left of the frame indicator.

2. Step through the simulation backward and forward, a frame at a time, using the arrows to the right of the frame indicator.

3. Drag the frame indicator to move to any particular point in the simulation.

Note that at the beginning of the simulation, the Backward Play button appears gray because you cannot run the simulation backward when it has not yet been run forward. Once you select one of the Play or Step buttons, the button changes its function: it

becomes a Pause button, as shown in Figure 1.16, and clicking on it pauses the simulation at any point. The number 50 in Figure 1.16 indicates the number of frames that have been completed so far in the simulation.

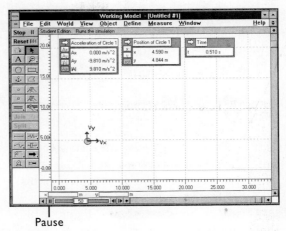

Pause

Figure 1.16

To play back the simulation:

- Select the **Forward Play** button on the tape player.

When the mass seems to be at its highest point (where the velocity component in the y direction is zero),

- Click on the **Pause** button of the tape player.

Now use the forward or backward step buttons (to the right of the frame indicator) and observe the y coordinate on the Position meter to locate the maximum height of the mass. This will occur at frame 122, when $y = h = 7.339$ m, and $t = 1.220$ sec. Figure 1.17 shows the mass at its maximum height. The y-coordinate velocity vector points up until frame 122, after which the value of y will be decreasing.

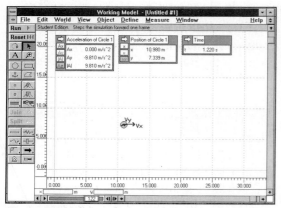

Figure 1.17

In the same manner, locate the distance and time when the mass crosses the x axis (i.e., when y is as close to zero as possible), in this case at frame 245, when $x = d = 22.050$ m, $y = -0.042$ m, and $t = 2.450$ sec. Figure 1.18 shows the mass crossing the x axis.

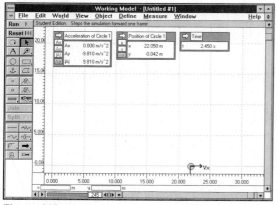

Figure 1.18

- Click on **Reset.**

The simulation is now complete.

Link to Dynamics

Stepping through the experiment with the Tape Player, note that the velocity in the vertical direction (shown by the vertical component of the velocity vector) was zero at the highest point (i.e., when the direction of the mass is reversing), and the acceleration at all times was the constant acceleration due to gravity (9.81 m/s^2 downwards).

Can you calculate the maximum height and the distance that the mass travels from your knowledge of dynamics?

How do your calculations compare to Working Model's numbers?

Using Air Resistance

This problem would be more realistic if the effects of air resistance were included. Air resistance in Working Model is treated as a force on a moving body opposite to the direction of its motion. This force is proportional to the body's cross sectional area in the direction of motion. Air resistance in Working Model does not take into account the coefficient of drag of various shapes. The default setting for Air Resistance in Working Model is none, but you can easily turn Air Resistance on.

- Select **World** menu, **Air Resistance...**
- Choose **Standard.**
- Click on **OK.**
- **Run** the simulation.
- Click on **Reset** when finished.

Compare the results with and without air resistance.

In later tutorials you will see how to use other descriptions for air resistance, but you can always quickly estimate its effects by using the Standard Air Resistance setting.

Saving a Simulation

You have finished your simulation and should save it for future use.

- Select **File** menu, **Save.**

The File Save As dialog box appears to alert you that the simulation has not yet been given a name. (Refer to Chapter 2, Working Model Basics, in Part 1, Getting Started, if you need to review the operation of this dialog box.)

W You must follow the same file-naming conventions necessary for DOS. Your file name cannot exceed eight characters. Working Model automatically adds the .WM suffix to all of your file names; you do not need to type it.

W Highlight the text in the box below **Filename:** (if necessary).

W Type **TUTOR#1**

W Click **OK.**

M You may name your file as you name all other files, with no conventions.

M Highlight the text in the box below **Save this document as:**.

M Type **Tutorial #1**

M Click **Save.**

Tip It is a good practice to use descriptive names for your files. If you name your simulation of a rocket ship blast-off ROCKET.WM or Rocket, you should have no problem calling it up again when you need it. On the other hand, if you have two rocket ship simulations, one of a blast-off and one of a landing, you might call your files BLASTOFF.WM or Rocket Blast-off and LANDING.WM or Rocket Landing, so that you can easily distinguish between them.

It is a good idea to save your file periodically as you work on a simulation, so that if something goes wrong, you will not have to start all over again.

Exiting Working Model

After you have saved your file, you can safely exit your Working Model session.

W Choose **File** menu, **Exit.**

M Choose **File** menu, **Quit**

1.1 The Basics   Set up the workspace with X,Y Axes on and create a circle on the horizontal axis. Use the Properties window to place the coordinates of the circle at 0 m in the y direction and at 5 m in the x direction. Scroll the view so the circle and axes appear in the lower-right-hand corner of the workspace. Give the circle an initial velocity of 10 m/s upward and 5 m/s to the left. Run the simulation. You may need to adjust the view with the Zoom tool, so that the whole simulation is visible in the workspace. Reset the simulation, add a time, position, and velocity meter for the x,y location of the circle, and run the simulation. (Create a velocity meter by selecting velocity from the same menu you used for the acceleration meter in the tutorial.) With the help of the Tape Player, find the maximum height and range that the circle travels and the time at which these maxima occur. Rerun the simulation with an acceleration meter and see if you can understand why the acceleration behaves as it does.

1.2 Applying the Basics   You are a football quarterback and also an engineer. You want to understand how you can throw the longest possible pass given an initial net velocity of 10 m/s. With the same setup as Exercise #1, run simulations with different initial vertical velocities and corresponding horizontal velocities that vectorially add to 10 m/s. Start with initial vertical velocities of 2, 5, and 8 m/s, then home in on an optimum. Record the maximum height and range and their time of occurrence for each simulation, and relate these to the initial velocities. Using principles from dynamics, calculate height and range based on the initial velocities and compare your results to those of the Working Model simulation.

1.3 For Fun and Further Challenge   You are an engineer for NASA working to avoid a catastrophe. Remnants of an asteroid are soon to plummet through the atmosphere, heading directly for Disney World! Your job is to intercept the asteroid high in the sky with a projectile, as shown in Figure E1.1, blowing it to little bits. Working with a scale model, locate the projectile at the origin of the workspace and the asteroid at position (0,10) m with an initial velocity of (5,–4) m/s. Your job is to give the projectile the correct initial velocity to intercept the asteroid, disintegrate it, and save the day. (Trial and error works well.) Add velocity meters, step through the simulation with the Tape Player, and see how the velocity changes during impact. Develop a method of calculating the initial conditions of the projectile to ensure that it intercepts the asteroid at an arbitrary point on the asteroid's path. Note: If you are successful, you win free tickets to the amusement park!

Figure E 1.1

1.4 Design and Analysis   In this exercise, you will be progressively developing a three-degree-of-freedom (planar) vehicle suspension model, and examining its dynamics, vibration, and control. As you become more sophisticated with Working Model in the course of future tutorials, you will add to its complexity, until the model is complete and you are hired by a big automotive firm. To start, turn off gravity by selecting **World** menu,

Gravity...; click on **None**; then click on **OK**. Create a "car" with the Circle tool (you have only made circles so far; you will develop the car model in later tutorials) and place it at the origin, as shown in Figure E1.2. Turn on Standard Air Resistance, give the car an initial horizontal velocity of 25 km/hr, and see how long it takes the car to coast to a stop. Determine appropriate meters to help yourself. Adjust the constant k in the Air Resistance dialog box to allow the car to coast 10 m more before stopping. Can you create a dynamic model that would predict the relationship between the level of wind resistance and the coasting distance?

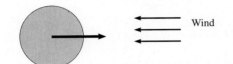

Figure E1.2

Exploring Your Homework   These textbook problems involve objects that can be modeled as point masses in plane motion and that move under the influence of gravity. You should convert the units in the problems to SI units before setting up the simulation and use Working Model's setting for High Air Resistance.

1.5   Engineers testing a vehicle that will be dropped by parachute estimate that its vertical velocity when it reaches the ground will be 20 ft/s.

If they drop the vehicle from the test rig in Figure E1.3, from what height (h) should they drop it to simulate the parachute drop?

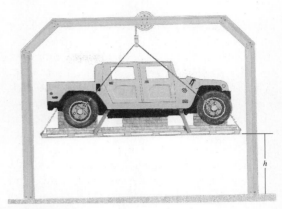

Figure E1.3

1.6   A 100-slug projectile is launched from $x = 0$, $y = 0$, with initial velocity $v_x = 400$ ft/s, $v_y = 400$ ft/s. (The y axis is positive upward.) The aerodynamic drag force is of magnitude $C|v|^2$, where C is constant. Determine the trajectory of the projectile for values of C equal to 0.002, 0.004, and 0.006.

Exercises 1.5 and 1.6 are adapted from Bedford and Fowler's Engineering Mechanics: Dynamics *(Addison-Wesley, 1995), pages 22, 130.*

Rectilinear Relative Motion

Objectives

In this tutorial you will learn to:
- Create multiple rectangles
- Select and deselect objects
- Change the size of objects
- Label objects
- Move objects in the workspace
- Select groups of objects and change their properties
- Use formulas
- Anchor objects
- Create and work with output meters

Dynamics principles related to this tutorial

- Rectilinear motion of a particle with constant acceleration
- Velocity of a particle as a function of time
- Relative position and velocity of two particles

Introduction

Relative motion problems are important in many areas of mechanical engineering, including vehicle guidance systems and machine design. In this tutorial, you will learn how to determine the relative motion of two objects. You will use Working Model to create and change the properties of multiple objects. You will also learn how to attach objects to a fixed reference frame, how to specify variable velocity for an object, and how to label objects.

Problem Statement

Two athletes are 10 m from the finish line of the Boston Marathon. One of them, athlete A, is moving at a pace of 4 m/s (14.4 km/hr or 8.95 mph). The other, athlete B, is moving at half that speed (2 m/s). Athlete A figures that if she can just maintain her speed she will win. Athlete B knows that to come in first she will need to speed up. She starts to accelerate at a rate of 2.5 m/s². You will create a Working Model experiment to see (1) which athlete will reach the finish line first, (2) how long it will take, and (3) what the relative position and velocity of the two athletes will be at the finish line. Model the athletes as two identical blocks on frictionless surfaces, as shown in Figure 2.1.

Figure 2.1

Block A

Block B

Starting Working Model

Begin Working Model by double-clicking on its icon. A new window, Untitled#1, opens. Resize it to fill your workspace. If you need help resizing your workspace, refer to Tutorial 1.

> **Tip** Sometimes the Maximize button is not visible when the document window opens. If this happens, click on the title bar of the document window and drag the window until the Maximize button is visible.

Your workspace should open with the settings you selected for X,Y Axes, Grid Lines, Rulers, and Grid Snap in Tutorial 1 already on. If you need help setting up your workspace, refer to Tutorial 1.

Creating the Components

The Rectangle icon looks like a rectangle and is found on the tool bar to the right of the Circle icon. The Rectangle icon has a small arrow on its right side; this means that holding the cursor down on that icon causes another icon to become available: the Square icon. The Rectangle tool is used to create rectangular bodies, and the Square tool is used to create square bodies.

- Double-click on the **Rectangle** icon on the tool bar to make it active.

- Click in any four locations in the workspace.

- Click on the **Arrow** icon on the tool bar to deactivate the Rectangle icon.

Your workspace should look like Figure 2.2. The rectangles look like squares, because you have not sized them yet. You can change their size and shape to make them look like rectan-

gles. If you had used the Square tool, you could only change the size of the squares; you would not be able to make them into rectangles.

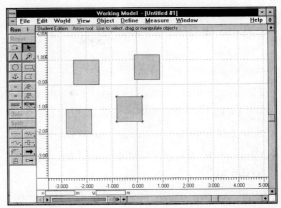

Figure 2.2

Changing the Size of Objects

You can change the size and shape of an object by clicking on and dragging it.

- Click on one of the **rectangles.**

Four small blocks appear at the corners of this object, showing that it has been selected, as you can see in Figure 2.2.

- Click and drag any of these **corner blocks** to change the size of the object.

There is a much more precise way to change the size of a body, which you will now use to alter the size of two of the rectangles so that they have identical dimensions and look like Blocks A and B in Figure 2.1.

Selecting and Deselecting Objects

As you learned in Tutorial 1, you select an object by clicking on it. You can use the Shift key in conjunction with clicking to select more than one object in the workspace. You can also use the Shift key in conjunction with clicking to remove an item from the current selection. You can select all of the objects in the workspace, using Edit menu, Select All.

- Click on one of the rectangles in the workspace.

- Holding **Shift** down, click on any other rectangle in the workspace.

- Select **Window** menu, **Geometry...**

- Change the **height** to **0.4** m and the **width** to **0.6** m.

- Click on one of the rectangles you have not yet changed.

- Holding **Shift** down, click on the other rectangle you have not yet changed.

- Change the **height** to **1** m and the **width** to **4** m.

- Close the **Geometry window.**

Labeling Objects

The Appearance window controls the appearance of objects. You will label the two small blocks as A and B, and the two long rectangles as ground, using the Appearance window. These labels will make it easier for you to follow the directions as you proceed through this tutorial.

- Click on one small rectangle to select it.

- Select **Window** menu, **Appearance...**

The Appearance window shown in Figure 2.3 appears.

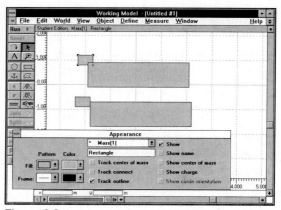

Figure 2.3

The Appearance Window

You can change the appearance of any Working Model object, such as its color, fill pattern, or name, by changing the information in the Appearance window. You can change the color and/or pattern of a body by using the Fill pull-down menus, and you can change the width and/or color of the line drawn around the body by using the Frame pull-down menus. (**M** Clicking on the boxes for Fill, Color, and Frame cause a menu of choices to pop-up.)

To display the Appearance window for one or more objects, select the objects and then choose from the Window menu, as you did above. All windows show data for the current object selection set. Changing the data in the windows changes the data for the currently selected object or objects.

In this tutorial you will use the Appearance window to label the objects in the workspace by changing their default names to more descriptive ones. You will learn how to use the other options in the Appearance window in later tutorials.

- Highlight the name Rectangle in the text box and change it to **A.**
- Click on **Show name** so that a check mark appears in the box to its left.
- Click on the other small rectangle.

The Appearance window now shows information for that rectangle.

- Change the name Rectangle to B.
- Click on **Show name** again.
- Click on one of the long rectangles representing the ground.
- Holding **Shift** down, click on the other long rectangle.

The Appearance window now shows information for the two ground rectangles.

- Change the name Rectangle to **ground.**
- Click on **Show name.**
- Close the **Appearance window.**

Moving Objects in the Workspace

If you click and drag an object (anywhere but on the corners of the object), you can move the object around within the workspace. As demonstrated in Tutorial 1, the Grid Snap option both helps to position objects with precision and hinders that process; you may need to re-select your point on the object, or turn Grid Snap on and off, to position the blocks where you want them.

You will use X,Y Axes to help you locate the left edge of each block at $x = 0$. Refer to Figure 2.4 as a guide.

Tip Try not to click and drag on the center of an object, as this will create a vector. Avoid clicking on the square handles at the corners, too, as this resizes the object.

W Click and hold down the mouse button over an object until the cursor changes to a cross made of four arrows. This indicates that dragging will move the object, not resize it.

- Click on and drag all four rectangles and arrange their left edges at **x = 0.**

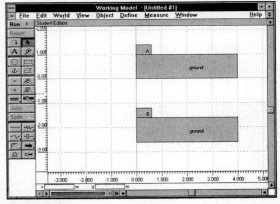

Figure 2.4

Changing the properties of a group of objects

You can select an entire group of objects or a few at a time. To select them all,

- Select **Edit** menu, **Select All.**

Small corner blocks now appear on all the objects, as shown in Figure 2.5.

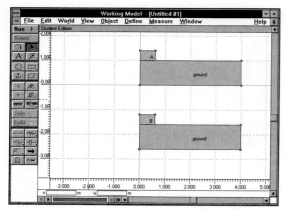

Figure 2.5

- Click and drag the objects so that the centers of **Blocks A** and **B** are approximately aligned with the position **x = 0.**

Your workspace should look like Figure 2.6.

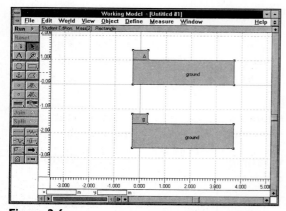

Figure 2.6

- Double-click on any object to open the **Properties window.**

The words "mixed selection" appear just below the title bar in the Properties Window; this is because you have more than one item selected.

- Click on the down arrow next to **mixed selection.**

A list of all the objects in the workspace appears. Note that dashes appear to the left of all of the objects in the list because they are all included in the mixed selection. Whenever "mixed selection" appears in a window, you can use this pull-down menu to see which items in your experiment are included in the mixed selection.

- Highlight the text box to the right of **stat. fric.**

- Change to **0.**

- Highlight the text box to the right of **kin. fric.**

- Change to **0.**

Tip Pressing the Tab key is an easy way to cycle through input boxes and highlight the text within them. Pressing Shift + Tab cycles through in reverse.

This ensures that all sliding surfaces will be frictionless. Refer to Figure 2.7.

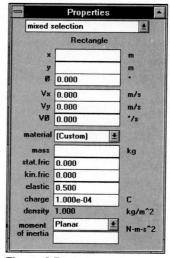

Figure 2.7

Now you want only Blocks A and B to be selected.

- Hold the **Shift** key down and click on the two long rectangles that represent the **ground.**

They are deselected and only Blocks A and B are shown as selected. Refer to Figure 2.8.

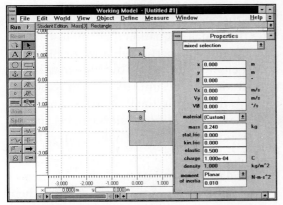

Figure 2.8

- Change the text box to the right of **x** to **0** m.

This places the centers of both blocks precisely at $x = 0$. Since your blocks were moved with Grid Snap on, they were probably already centered at $x = 0$.

- Holding down the **Shift** key, click on **Block B.**

Only Block A is selected now. You now want to give Blocks A and B their initial velocities and accelerations.

- Highlight the text to the right of **V_x.**
- Change to **4** m/s.
- Press **Enter.**

An arrow appears on Block A, showing this velocity. Refer to Figure 2.9.

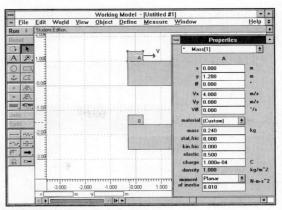

Figure 2.9

Using Formulas

You can use formulas in place of numbers in any field of a Properties window. This means that you can attach equations to govern the motion as well as the physical properties of masses during a simulation. For example, you could use formulas to model the mass of a rocket that becomes lighter as its fuel is burned.

Formulas follow standard rules of mathematical syntax, and strongly resemble the equations used in spreadsheets and programming languages. See Appendix B for more on the formula language used by Working Model.

- Click on **Block B.**

According to the Problem Statement, this mass has an initial velocity of 2 m/s and an acceleration of 2.5 m/s^2. The equation representing the velocity of this point is $v_x = a_x t + v_{x0}$, where a_x = acceleration = 2.5 m/s^2, t = time, and v_{x0} = initial velocity = 2 m/s. This can be entered in the V_x location, just as the constant velocity was for Block A.

- Highlight the text to the right of V_x.
- Enter **2.5*time+2**
- Close the **Properties** window.

Anchoring Objects

The Anchor tool can be used as a position or velocity constraint. You will use the Anchor tool as a position constraint on the ground rectangles, and as a velocity constraint on Block B. Anchors are activated by placing them on masses.

- Double-click on the **Anchor** icon on the tool bar (it looks like an anchor).
- Click once on each of the rectangles that represent the **ground.**
- Click once on **Block B.**
- Click on the **Arrow** icon to deactivate the Anchor tool.

Your workspace should now look like Figure 2.10.

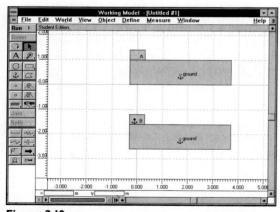

Figure 2.10

In Tutorial 1 you learned that the values in the Properties window represent initial values.

Once a simulation is started, the object starts from these initial conditions and moves according to the Working Model settings governing the simulation (such as gravity, air resistance, and so on). When an anchor is placed on an object, it allows the values in the Properties window to constrain the object's movement throughout the simulation. The anchor tool can constrain velocity or position.

When the anchor is placed on Block B, it activates the velocity equation that you just entered in the Properties window for Block B. Its velocity will be defined by this equation for the duration of the simulation. In the case of Block A, the value in the Properties window is just the initial condition (although, because there is no friction to slow it down, it will remain the same).

In a similar way, placing anchors on the two rectangles that represent the ground fixes their location based on the location defined in the Properties window. Because the values for the position of these rectangles are constants, the anchors act as positional constraints. If there were a continuously changing formula in their Properties window, the anchors would cause them to move as so defined.

In general, if there is no formula inserted for either position or velocity, then the anchor fixes the body in the workspace.

- Click and drag outward on the **Zoom** icon while you hold the mouse button down.
- Select the **Zoom Out** icon (the magnifying glass with the minus sign in it) by releasing the mouse button when it is highlighted.
- Click once on the workspace.
- Click on the **Arrow** icon to deactivate Zoom.

- Move your view to the left side of the workspace. If you need help, refer to Tutorial 1.

- Enlarge each of the ground rectangles by clicking and dragging a corner of the rectangle so that it extends past the **13 m** mark on the horizontal ruler.

Your screen should now look like Figure 2.11.

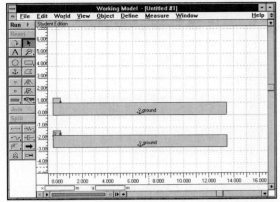

Figure 2.11

Running the Simulation

- Run the simulation now; if you need help, refer to Tutorial 1.

You will see that Block A is initially ahead of Block B, but that Block B eventually catches up with and passes Block A.

- **Reset** your simulation.

Obtaining Results

You will create output meters to help you determine the numerical results of your simulation.

- Select **Measure** menu, **Time.**

- Select **Block A.**

- Select **Measure** menu, **Position, X Graph.**

- Select **Block B.**

- Select **Measure** menu, **Position, X Graph.**

When clicking on Block B, be sure to select the block and not the anchor.

> *Tip* At the top of your screen, there is a gray line called the Help Ribbon. The Help Ribbon gives a concise description of the tool or object located at the mouse cursor. If you are positioned over Block B, the Help Ribbon will say Mass[3] Rectangle; if you are positioned over the anchor, the Help Ribbon will say Point[7] Anchor: Prevents the mass it is attached to from moving. (Your mass numbers may be different.) Selecting View menu, Workspace, Help Ribbon shows or hides the Help Ribbon.

Position the meters conveniently. If you need help repositioning them, refer to Tutorial 1.

Your workspace should look like Figure 2.12.

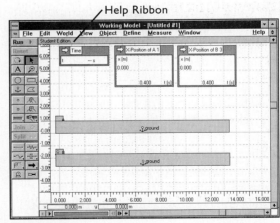

Figure 2.12

Changing the Meter Display

Output meters can give information in three forms: numerical/digital, line graph (referred to as graph), or bar graph (referred to as bar). The meter display is controlled by an arrow in the upper-left corner of the meter.

W In Windows it is an arrow pointing to the right, as shown in Figure 2.13. You toggle the display by clicking on the arrow. The display always changes in this order: digital → graph → bar.

W **Figure 2.13**

M On the Macintosh it is a down arrow, as shown in Figure 2.14. A pull-down menu with the options Digital, Graph, and Bar appears when you click on and hold the arrow.

M **Figure 2.14**

W Click twice on the arrow at the upper-left-hand corner of each position meter to obtain the **Digital** display. (Time is already digital.)

M Click on and drag the arrow at the upper-left-hand corner of each position meter to select **Digital** as the display mode.

Note: In the remainder of the tutorials, you will simply be instructed to change the display to Digital, Graph, or Bar; the different steps for Windows and Macintosh will not be provided.

- Select the **Position** meter for **Block A.**
- Select **Window** menu, **Appearance...**
- Highlight the text **X-Position of A 1**. (Your meter may have a different number next to the letter A, depending on the order in which you created your blocks.)
- Change the name on the Position meter to read **X-Position, Block A.**
- Click on the **Position** meter for **Block B.**

You may need to move the Appearance window so that you can see and select all the elements on the screen.

- Highlight the text **X-Position of B 3**. (Your meter may have a different number next to the letter B, depending on the order in which you created your blocks.)
- Change the name of the Position meter for Block B to read **X-Position, Block B.**

Your workspace should look like Figure 2.15.

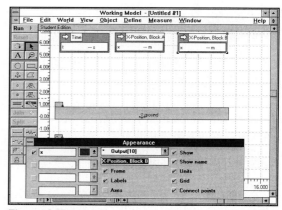

Figure 2.15

Measuring Relative Position

You will now create a more complex meter to measure the position of Block A relative to the position of Block B. First you will create another X-position meter for Block A.

- Select **Block A.**

- Select **Measure** menu, **Position, X Graph.**

- Change the display to **Digital.**

- Move this meter to the lower-left-hand corner of your workspace. You may need to move the Appearance window slightly to place the meter.

Your workspace should be arranged as shown in Figure 2.16.

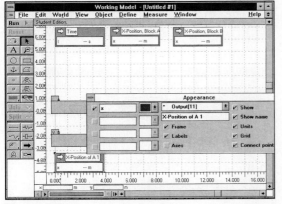

Figure 2.16

- Double-click on this new Position meter to open its Properties window.

You want this meter to show the position of Block A relative to Block B, i.e., $x_{A/B} = x_A - x_B$. Several changes need to be made to accomplish this. You will now make note of the numbers Working Model has assigned to Blocks A and B in your experiment so that

you can develop the formulas you will need for this tutorial.

- Select **Block A.**

- Write down the **mass number** shown just below the title bar of the Properties window here: **Block A = Mass[_____].**

- Select **Block B.**

- Write down the **mass number** shown just below the title bar of the Properties window here: **Block B = Mass[_____].**

- Select the **Position meter** you just created.

Note that the second line of properties in the Properties window, labeled y1, shows two quantities: x, the variable name or label, and mass[1].p.x, the Working Model variable formula or equation. The term mass[1] indicates the object of interest, and p.x represents the x component of the position. In Figure 2.17, mass[1] is Block A and mass[2] is Block B (if yours are different, use the correct numbers you wrote down).

Tip Don't enter any spaces in your object names (identifiers).

- Highlight **x,** next to y1.

- Change to **xr** (for relative position). This will change the output label on the meter.

- Press **Tab** to move to the next column.

- Enter **mass[1].p.x-mass[2].p.x** (remember to use the correct numbers for your Blocks A and B).

W You can make the Properties window wider by clicking the Maximize button in the top right corner, or by dragging on the right or left edge of the window when a double-ended arrow appears. This is helpful when entering longer equations.

M You can make the Properties window wider by clicking and dragging the resize box on the lower right corner of the window. This is helpful when entering longer equations.

If you enlarge your Properties window, it will look like Figure 2.17.

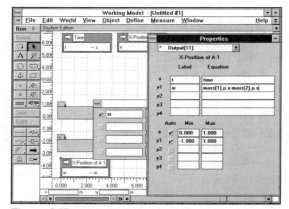

Figure 2.17

- Select the **Appearance** window again.
- Highlight the name **X-Position of A 1**. (Remember, your number may be different.)
- Change the name to **XA rel. to XB.**

Now you will create an x-velocity meter for Block A and modify it as you did the x-position meter to show velocity relative to Block B. The velocity of Block A relative to Block B is $v_{A/B} = v_A - v_B$.

Tip You may need to move or close and reopen the Appearance window to position the new meter.

- Select **Block A.**
- Select **Measure** menu, **Velocity, X Graph.**
- Change the display to **Digital.**

- Position the new meter to the right of the **XA rel. to XB meter.**
- In the Properties window for the new meter, change the label or variable name, **Vx,** to **Vr** (for relative velocity).
- Edit the Working Model variable formula to read **mass[1].v.x-mass[2].v.x** (remember to use the numbers that you wrote down for Blocks A and B).
- In the Appearance window, change the name from **Velocity of A 1** to **VA rel. to VB.**
- Close the **Appearance** and **Properties** windows.
- **Run** the experiment again.

Replay your experiment, using the Tape Player, to see which block will reach $x = 10$ m first and how long it will take. If necessary, refer to Tutorial 1 to review using the Tape Player controls.

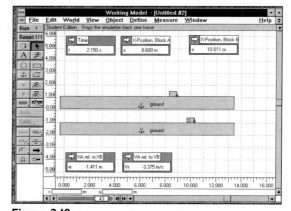

Figure 2.18

Figure 2.18 shows that at time $t = 2.150$ s, $x_B = 10.011$ m and $x_A = 8.600$ m. At this time, the relative position of Block A with respect to Block B will be $x_{A/B} = -1.411$ m (i.e., $x_A - x_B$), and the relative velocity will be $v_{A/B} = -3.375$ m/s.

<table>
<tr><td>

Link to Dynamics

</td><td>

The velocity equation that you inserted in the Properties window to govern the motion of Block B, $v_x = a_x t + v_{x0}$, can be integrated with respect to time to obtain Block B's position at any time, t. The resulting equation for the position of Block B is $x_B = 1/2 a_x t^2 + v_{x0} t + x_0$. The position of Block A can be obtained from an analogous equation

</td></tr>
</table>

with Block A having zero acceleration. Use these equations to validate the answers you obtained in your Working Model simulation. For this problem, you should have obtained answers with Working Model that are within 1% of the exact answers. The question of accuracy will be treated in later tutorials.

You may save this simulation as Tutor #2 (or Marathon) if you like. Exit Working Model. Refer to Tutorial 1 if you need help saving your simulation and exiting Working Model.

Tutorial 2 • Exercises

2.1 The Basics   Set up the workspace with X,Y Axes, Grid Lines, and Rulers. With the rulers as a guide, draw a 6-m by 0.2-m anchored rectangle to represent the ground. Create a small 0.4-m square block on the left end of the ground and a small 0.4-m diameter circle above the block in the "air." Your simulation should look like Figure E2.1. Use the Properties window for both the block and the circle to make sure that they have the same horizontal position. Give both the circle and block an initial horizontal velocity of about 4 m/s. Run the simulation. Change the block's properties to start with a 1 m/s initial velocity, but with a 2 m/s² acceleration. (Set Vx equal to **1 + 2 * time** in the Properties window.) Set up position meters for the block and the circle and give the meters more descriptive titles. Run the simulation again. Zoom in or out as necessary to see the whole simulation.

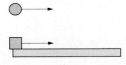

Figure E2.1

2.2 Applying the Basics   You are a technical consultant for a stunt team. Bill Biker wants to perform a stunt where he accelerates his motorcycle towards a gorge, hits his brakes too late, and skids over the edge and into the gorge. What the cameras won't show is that his velocity into the gorge is enough to carry him across to a landing site on the other side. The gorge is 30 m wide, the far side is 10 m below the near side, and there is 20 m of skid space before the gorge. A sketch of the setup for the stunt might look like Figure E2.2. Make the mass of Bill plus the bike equal to 300 kg in the Properties window. What must the velocity of the motorbike be just before the skid to keep Bill from certain death? Use Working Model's default friction to represent the skid. Use output meters and vectors to help illustrate your results. Show the vertical and horizontal velocities during this daring stunt.

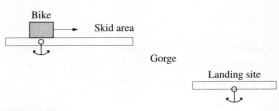

Figure E2.2

2.3 For Fun and Further Challenge
As an expert pinball athlete, you are in the national championships. At this point in the game, you have control over the direction and velocity the ball is about to take. As shown in Figure E2.3, if you can get the ball to exit to the right, you collect 10,000 points, but an exit to the left collects only 1,000 points. Set up the pinball game using the dimensions shown for the bumpers and gateways. By setting initial conditions, see what kind of champion you are.

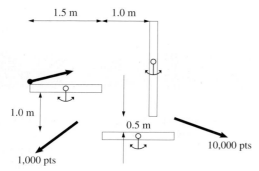

1.5 m 1.0 m
1.0 m
0.5 m
10,000 pts
1,000 pts

Figure E2.3

2.4 Design and Analysis
In the design of your suspension model, shown in Figure E2.4, you need to understand the effects of air pressure in tires. Experiment with a tire rolling down the road and hitting a bump. Use an effective wheel mass of 36 kg, a wheel radius of .3 m, and a velocity of 5 m/s. The bump is 0.1 m high. See how changing the elastic parameter of the wheel (air pressure) in the Properties window affects the dynamics of bounce. How does elasticity affect the height of the bounce and the velocity after the bump?

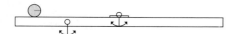

Figure E2.4

Exploring Your Homework
The following textbook problems have to do with accelerated motion in a straight line, with and without the influence of gravity. Most require setting formulas for positions, velocities, or acceleration, as was done in this tutorial. You can enter positions as functions of time in the Properties window just as you did for velocity in the tutorial. You will need to do some calculus to change accelerations to velocities ($a\,dt = dv$).

2.5 Suppose that the acceleration of a train during the interval of time from $t = 2$ s to $t = 4$ s is $a = 2t$ m/s^2, and at $t = 2$ s, its velocity is $v = 180$ km/hr. What is the train's velocity at $t = 4$ s, and what is its displacement (change in position) from $t = 2$ s to $t = 4$ s?

2.6 The position of a point is $s = 2t^2 - 10$ ft.

a. What is the displacement of the point from $t = 0$ to $t = 4$ s?

b. What are the velocity and acceleration at $t = 0$?

c. What are the velocity and acceleration at $t = 4$ s?

2.7 A rocket starts from rest and travels straight up. Its height above the ground is measured by radar from $t = 0$ to $t = 4$ s and is found to be approximated by the function $s = 10t^2$ m.

a. What is the displacement during this interval of time?

b. What is the velocity at $t = 4$ s?

c. What is the acceleration during the first 4 s?

2.8 The position of a point during the interval of time from $t = 0$ to $t = 6$ s is $s = \frac{1}{2}t^3 + 6t^2 + 4t$ m.

a. What is the displacement of the point during this interval of time?

b. What is the maximum velocity during this interval of time, and at what time does it occur?

c. What is the acceleration when the velocity is at maximum?

2.9 In a test of a prototype car, the driver starts the car from rest at $t = 0$, accelerates, and then applies the brakes. Engineers measuring the position of the car find that from $t = 0$ to $t = 18$ s, it is approximated by $s = 5t^2 + \frac{1}{3}t^3 - \frac{1}{50}t^4$.

a. What is the maximum velocity, and at what time does it occur?

b. What is the maximum acceleration, and at what time does it occur?

2.10 The velocity of an object is $v = 200-2t^2$ m/s. When $t = 3$ s, its position is $s = 600$ m. What are the position and acceleration of the object at $t = 6$ s?

2.11 The velocity of an object is given by $v^2 = k/s$, where k is a constant. If $v = 4$ m/s and $s = 4$ m at $t = 0$, determine the constant, k, and the velocity at $t = 2$ s.

Exercises 2.5–2.11 are adapted from Bedford and Fowler's Engineering Mechanics: Dynamics (Addison-Wesley, 1995), pages 26–29, 38.

Pulley Problem

Objectives

In this tutorial you will learn to:
- Create a pulley system
- Create and scale forces
- Define controls
- Use the Preferences dialog box
- Run experiments in Player mode

Dynamics principles related to this tutorial

- Newton's laws of motion
- Pulley systems
- Mechanical advantage

Introduction

Pulleys are widely used in engineering applications for transferring force and motion between bodies. They are used in the design of elevators and scaffolding, and in problems where it is necessary to input a small force through a large displacement in order to output a large force through a small displacement, or vice versa. Pulley problems illustrate the basic laws of motion quite nicely. Newton's first law states that a body at rest (or moving with uniform motion in a straight line), tends to stay at rest (or continue with uniform motion in a straight line), unless acted upon by an outside force. Newton's second law describes the accelerations that result when an outside force acts on a body. In this tutorial, you will use Working Model to solve a problem in which a pulley is acted upon by outside forces (gravity and an applied force), and you will evaluate the accelerations resulting from these forces. You will also learn how to create and use pulley systems with Working Model.

Problem Statement

The ideal pulley system shown in Figure 3.1 might represent a scaffolding system. Mass A would represent the scaffold and mass B would be where the rope could be wound or unwound by a motor to raise or lower the scaffold at a reasonable speed. Determine the tension in the cable and the acceleration of the two bodies for three cases: (1) $F = 0$, (2) $F = 4$ N, and (3) $F = 10$ N, where F is the force on the body. Assume that the pulleys are massless and frictionless.

Figure 3.1

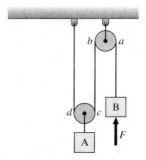

Starting Working Model

Open Working Model by double-clicking on its icon. Resize your window to fill the screen (if necessary) before you begin.

Creating the Components

You will create two rectangles. Align the rectangles with the grid lines, as shown in Figure 3.2. Grid Snap and the coordinates will help with the alignment.

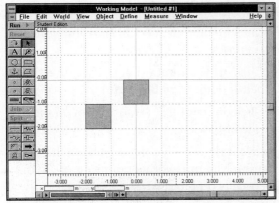

Figure 3.2

- Double-click on the **Rectangle** icon.
- Place the **center** of one rectangle at (**-1.5**, **-1.5**) m.
- Place the **center** of the second rectangle at (**0, -0.5**) m.
- Click on the **Arrow** icon to deactivate Rectangle.

You will now label the rectangles, using the Appearance option on the Window menu. If you need help with labeling objects, refer to Tutorial 2.

- Click on the **left-hand block.**
- Use the Appearance window to change its name from Rectangle to **A.** (Remember to select **Show Name**.)

Your Appearance window should look like Figure 3.3. Now use the same method to label Block B.

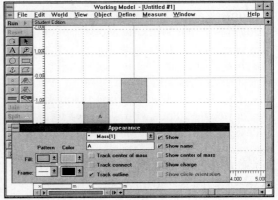

Figure 3.3

- Change the name of the **right-hand block** to **B.**
- Close the **Appearance** window.

Creating a Pulley System

Pulleys in Working Model behave as a single rope going through multiple fixed points. A pulley tool creates a rope and several pulleys. You define the starting point of the rope with a single click in the workspace. Then you define each pulley with a single click. Double-click to signal the endpoint.

Any individual pulley can be attached to either the background or a mass object. The total length of the rope is fixed, but the partial length between each pair of points can vary.

Pulleys in Working Model are somewhat different from the pulleys shown in Figure 3.1; because of this, the Working Model simulation will look different from Figure 3.1.

Real pulleys have a shape. Working Model pulleys consist of two endpoints, multiple intermediate points, and a rope connecting them. The tension in the rope is always uniform. Working Model pulleys are massless and frictionless, as are the pulleys for this problem. Recognizing that Working Model pulleys have zero size, you must arrange them so that the size shown in Figure 3.1 is imitated. This is easiest to understand if you think of Working Model pulleys as threads going through the eyes of needles. Thus, you will need to place pulleys as points *a*, *b*, *c*, and *d*, as shown in Figure 3.1. This will locate all the Working Model rope ends exactly where they need to be to solve the problem.

You can define pulley length and elasticity in the pulley's Properties window. Length is the actual length of the pulley system, and Current Length is the length of a line connecting each point in the pulley system (if the pulley is slack, Length is greater than Current Length). You will learn more about working with elasticity when you work with ropes in Tutorial 6.

- Click on **Block B.**

A small circle appears at the *mass center* of the block. You must locate the first point of the pulley system exactly at the mass center, or the block will rotate during its motion.

- Click on the **Pulley** icon on the tool bar (second icon from the bottom on the left).

Tip As you position the cursor over the icons on the tool bar, the name of each icon appears in the Help Ribbon, just below the menu options.

- Click on the **mass center** of Block B.

Grid Snap will help you place the first pulley point accurately if you have located the rectangles carefully. Refer to Figure 3.4 to select the rest of the points for the pulley system.

- Click on point **a** at **(0, 2)** m directly above the center of Block B.
- Click on point **b** at **(-1.2, 2)** m above the right side of Block A.
- Click on point **c** at **(-1.2, -1.1)** m directly below point *b*, on Block A.
- Click on point **d** at **(-1.8, -1.1)** m on the left side of Block A.

To avoid rigid-body rotation during the motion, you must locate points *c* and *d* symmetrically on Block A.

- Double-click on the final point, point **e**, at **(-1.8, 2)** m directly above point *d*.

This completes the pulley system. Your system should look like that in Figure 3.4.

Now use the Appearance window to label points *a* through *e*. Note that you do not want to label the mass center of Block B. Refer to Figure 3.4 to label the points correctly.

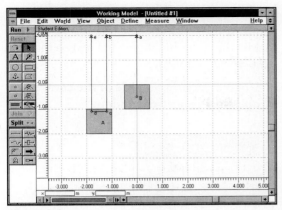

Figure 3.4

- Click anywhere in the workspace to deselect the pulley system, then select point *a*.

- Open the **Appearance** window.

- Change the name **Point** to **a** and select **Show name.**

- Use the pull-down menu of entities in the Appearance window to select the other points in turn, changing their names to match Figure 3.4.

- Close the **Appearance** window when your labels match Figure 3.4.

Creating and Scaling Forces

In Working Model, a *force* acts on the body to which it is attached. The point of application can be anywhere on the body. Forces are attached to the body lying under the cursor when you click on it.

- Click on the **Force** icon on the tool bar (to the right of the Pulley icon).

- Click and drag from one Grid Snap below the center of **Block B** down about 1 m.

Use the Coordinates display to help you draw the force so that your screen looks like Figure 3.5. To move a force, you can click on it anywhere except its endpoint, and drag it to a new location. If you drag on a force's endpoint, you will change its size, not its position.

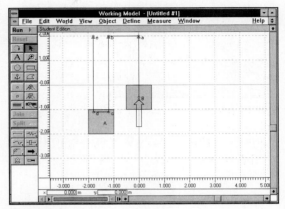

Figure 3.5

- Double-click on the **force** to open its **Properties** window.

The Force Properties window is different from those you've seen in earlier tutorials. As shown in Figure 3.6, using the Properties window, you can (1) define the force precisely in terms of rectangular (Cartesian) or polar coordinates, (2) make it rotate with the mass (the default is not to rotate), (3) identify its base point, and (4) define conditions under which it will be inactive. You will learn more about working with the Active when option in Tutorial 8.

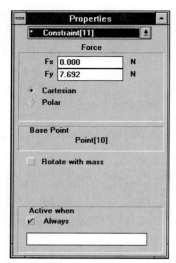

Figure 3.6

- Set the values of $\mathbf{F_x}$ to **0** N, and $\mathbf{F_y}$ to **10** N, if they are not already at those values.

- Close the **Properties** window.

Defining Controls

Controls allow you to adjust simulation parameters before and *while* a simulation is running. For example, you can control the magnitude and direction of a force directly from the workspace. A control can be a slider (the Working Model default), a text box, or a button. In Tutorial 6 you will see how to use controls to switch between simulations. In this problem, you want to observe the system for several different values of force. To make changes in force easier, you will choose to control the magnitude of the force by defining a Working Model control.

- Click on the **force.**

- Select **Define** menu, **New Control, Y Force.**

Clicking on the slider control changes the value of the force, and clicking on the text box below the slider highlights the text so that you can type a new value.

- Double-click on the **title of the control** to open its **Properties** window.

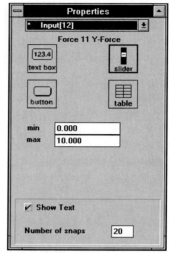

Figure 3.7

The Control Properties window is different from those you've seen before. As you can see in Figure 3.7, you can change the type of control to be a slider, text box, or button, or you can use an external table file to feed the data into the control.

You can specify minimum and maximum values for a *slider* control (these values should now be 0.0 and 10.0, respectively). You can alter these if you would like the control to operate over a different range of values. Number of snaps indicates how many discrete values are available in the range of the slider. By default, a text box is attached to the slider control. You can use the text box to enter a precise value within the range, even if that value does not coincide with the discrete steps of the slider. You can turn this text box on and off by clicking on the Show Text check box.

A *text box* control allows you to enter a precise numerical input of the property value.

Using a *button* control, you can select one of the two values specified in the min and max boxes. You can use the button as a toggle switch or as a press-and-hold button.

A *table* control reads its values from a *table file* (an ASCII text file containing columns of numbers); Working Model assumes that the first column of data is for *time* and the second is for the control values. Using this feature, you can combine simulational data with your simulations. When you select the table button, you will be prompted for the ASCII file containing your input data.

In addition, Working Model is capable of communicating with another application in real time by exchanging data through controls. See Appendix C for more about real-time links with external applications.

For this simulation you will leave the control set to slider and close the Properties window to accept all of its default settings.

- Close the **Properties window**.

You can now vary the force by moving the slider on the control in the workspace. You can move the slider by clicking at any location on it, by clicking on it and dragging, or by entering a value in the text box.

Running the Simulation

Set the Y force to different values and observe how the blocks move.

- **Run** the simulation.
- **Reset** the simulation.

Obtaining Results

Create and properly label output meters with digital results for the tension in the cable and the acceleration in the y direction of the two blocks. If you need help creating output meters, refer to Tutorial 2.

- Select the **pulley system**.
- Select **Measure** menu, **Tension**.
- Reposition and rename the meter to match Figure 3.8.
- Select **Block A**.
- Select **Measure** menu, **Acceleration, Y Graph**.
- Change the meter to a **digital** display.
- Reposition and rename the meter to match Figure 3.8.
- Repeat for **Block B**.

When you are finished, your workspace should look like Figure 3.8.

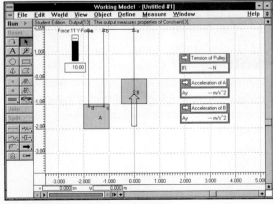

Figure 3.8

- **Run** the simulation again for the three force values, 0, 4, and 10 N.

You should obtain the following results:

F (N)	T (N)	a_A (m/s^2)	a_B (m/s^2)
0.00	5.886	1.962	−3.924
4.00	5.086	0.362	−0.724
10.00	3.886	−2.038	4.076

Note that the acceleration of Block B is always twice the magnitude of the acceleration of Block A. If you were to examine the distances traveled by the two blocks, you would see that Block B also travels twice the distance that Block A travels for any given time period. Thus, this pulley system creates a "mechanical advantage," where one block is moved a certain distance, and the other block moves twice that distance.

Link to Dynamics

The relationships governing the motion of pulley systems can be derived from simple kinematic relationships. For example, the motion of the present system can be described by one equation which defines the rope as having a constant length:

$2 y_A + y_B = $ constant

Differentiating this equation yields:

$2 v_A + v_B = 0$

Differentiating again yields:

$2 a_A + a_B = 0$

These equations show that Block B will always move twice the distance of Block A (in the opposite direction), and that the velocity and acceleration of Block B will always be twice the velocity and acceleration, respectively, of Block A. Open position and velocity meters for the two masses to demonstrate that the relationships for position and velocity are consistent with the equations above.

- **Save** your simulation as **TUTOR#3** or **Tutorial #3**.

Using the Preferences Dialog Box

You used the Preferences dialog box briefly in Tutorial 1 to save your workspace settings for use in future files. The Preferences dialog box gives you control over several important run-time features that you will want to know about. You will open a new Working Model window to simulation with these.

Note: You do not need to close one file to open another or to open a new file. Your new file opens on top of your current file and you can toggle between them.

- Select **File** menu, **New.**

A new window, Untitled#2, opens on top of your TUTOR#3.WM or Tutorial #3 file.

- Resize the workspace if necessary.

- Create a rectangle anywhere in the new workspace.

- Click and immediately drag the rectangle to another location; continue to hold the cursor down.

Note that the rectangle remains in its original location and an outline of the rectangle moves with the cursor, as shown in Figure 3.9.

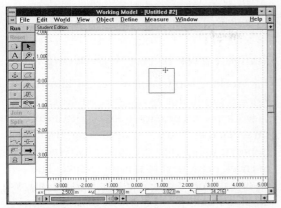

Figure 3.9

- Release the cursor.

The rectangle moves to the new location.

- Click on the rectangle again, and hold the cursor down, but now wait a second before dragging.

Note that the rectangle outline no longer remains in the workspace. You can change this behavior in the Preferences dialog box so that there will never be an outline when you click and drag.

- Select **World** menu, **Preferences...**

The dialog box shown in Figure 3.10 shows a list of options and their default settings. The current setting for editing objects is to edit them as *outlines*, which is faster than editing them as *objects*. You will change this setting now.

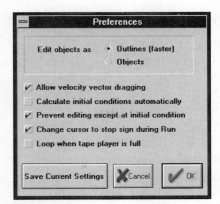

Figure 3.10

- Click on the button for Edit objects as **Objects.**

- Click **OK.**

- Click and drag the rectangle to another location; continue to hold the cursor down.

The object should move along with the cursor; the outline no longer appears. You will now examine the other options available in the Preferences dialog box.

- Select **World** menu, **Preferences...** again.

The other options in the Preferences dialog box are:

Allow velocity vector dragging

In Tutorial 1, you gave the soccer ball an initial velocity by dragging on its mass center and creating a velocity vector. You can only do this if the Allow velocity vector dragging option is on; otherwise you must use the Properties window to specify initial velocities for that object, and must use the Define menu, Vectors selection to show vectors.

Calculate initial conditions automatically

Working Model normally does no computation before you run a simulation. This means that meters and graphs are normally blank at time $t = 0$. After you run and reset a simulation, you will see the proper values in meters at $t = 0$. Working Model can automatically calculate the results at $t = 0$ after every editing operation. The drawback to this is that the calculation takes time. For complex simulations, this can produce a time delay between editing operations.

The advantage of automatic initial conditions calculation is that meters and vectors are immediately displayed with the proper values at all times. This is especially useful for statics problems, where vectors are being used to show forces. You should turn this option on if you want the proper force vectors to be displayed after every small adjustment to the simulation.

Prevent editing except at initial conditions

In these first three tutorials, you may have gotten the warning message, "Stop: You must reset to initial conditions before making changes," with the options Cancel and Reset. Choosing Cancel returned you to your simulation as it was. Choosing Reset returned you to the screen and reset your simulation to its initial conditions; you could then make any edits you chose. This is the Working Model default setting, which forces you to reset your simulation to $t = 0$ before making any changes. By turning this option off, you *can* make such changes in the middle of a simulation. Whenever you make any changes, however, you lose your old initial conditions. If you make only one change during the simula-

tion, you can use Undo on the Edit menu to undo the change and restore the previous conditions. If you make more than one change, however, you will not be able to reset to the original conditions; you can only Undo the very last change made.

Change cursor to stop sign during Run

This Working Model default setting allows you to stop a simulation by clicking anywhere in the workspace. The cursor changes to a stop sign while the simulation is running. You can disable stopping with the cursor by removing the check mark next to this preference. The cursor will not change to a stop sign when you run the simulation; you will need to use the Stop button on the tool bar or the pause on the Tape Player controls to stop the simulation.

Loop when Tape Player is full

By default, Working Model stops a simulation when there is not enough memory available to store further frames in the Tape Player. However, you can loop when the Tape Player becomes full. Additional frames are computed and stored on top of currently stored frames. The simulation will run until you stop it. Resetting in this situation will still return you to the initial conditions.

Save current settings

In Tutorial 1 you used this to save your workspace settings. By using this option, you can save a variety of document settings:

1. The size and location of the current document

2. All World menu settings (including the ones in the Preferences dialog box)

3. All View menu settings

4. Vector Length/Display settings on the Define menu

Do not save any settings you may change as you experiment with the Preferences here.

W If you select Save Current Settings, a file named WMPREFS.WM is created in your Windows directory. This file contains your settings and will be opened and read each time you open a new Working Model document.

M If you select Save Current Settings, a Working Model 2.0 Prefs file is created in the Preferences folder of your System folder. This file contains your settings and will be opened and read each time you open a new Working Model document.

> *Tip* If you want to delete the preferences you've saved and reset your Working Model to its original defaults, you can delete the WMPREFS.WM file in Windows or the Preferences file on the Macintosh.

You will now try some of these options in your pulley simulation. Working Model allows you to toggle between two or more simulations at one time. A list of the open files appears at the bottom of the Window menu options.

- Click on **OK**.

- Select **Window** menu, **TUTOR#3** or **Tutorial #3**.

The workspace you were working in earlier in this tutorial reappears.

- Change the **force** in the **control** to **4** N.

Note that the meters are all blank.

- Select **World** menu, **Preferences...**

- Click on **Calculate initial conditions automatically**.

- Click on **OK**.

The forces automatically appear in the meters, as shown in Figure 3.11; you do not even need to run the simulation to obtain the answers for this problem. You will now alter the preferences so that the cursor will not change to a stop sign while the simulation runs.

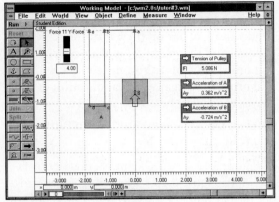

Figure 3.11

- Select **World** menu, **Preferences...** again.

- Click off **Change cursor to stop sign during Run**.

- Click **OK**.

- **Run** the simulation.

Note that a stop sign no longer appears as you move the cursor across the workspace.

- Select **Reset**.

If you like, you can continue to change the settings in the Preferences dialog box to observe the results of each change. It is most important that you keep in mind what options the Preferences dialog box offers so that you can use them to help you in future simulations.

Running Simulations in Player Mode

Until now you have been creating and running simulations in what is called Edit mode. This mode provides the full range of menus above the workspace and the tool bar to the left of the workspace. There is another option, called Player mode, which is meant for Working Model users who will be running simulations only as demonstrations, and will not be changing anything in them. In Player mode, the tool bar on the left of the screen is hidden, and the menu set is also reduced. You can insert controls to run the simulation. You will proceed with a typical application of the Player mode for the pulley problem.

- Select **Define** menu, **New Menu Button...**

The New Menu dialog box appears as shown in Figure 3.12. (The Macintosh dialog box is shown in Figure 3.13.) You can select a number of menu operations from this box, but you can create only one new menu button at a time. Creating menu buttons will be discussed more thoroughly in Tutorial 6.

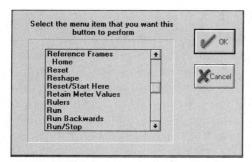

W   Figure 3.12

W   Scroll down the menu items until you find **Run**.

W   Select **Run** from the list.

W   Click **OK**.

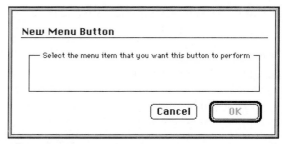

M   Figure 3.13

M   Select **World** menu, **Run**.

M   Click **OK**.

A button containing the word Run appears in your workspace. Next you will create a Reset button.

- Select **Define** menu, **New Menu Button...** again.

W   Scroll down the menu items until you find **Reset**.

W   Select **Reset** from the menu list.

W   Click **OK**.

M   Select **World** menu, **Reset**.

M   Click **OK**.

Move the buttons so that they are clearly visible in the workspace, as shown in Figure 3.14.

> **Tip** To reposition the menu buttons, click slightly outside the outline of the button so the handles appear. If you click on the button, you will run or reset the simulation. If this happens stop the simulation and try selecting the menu button again.

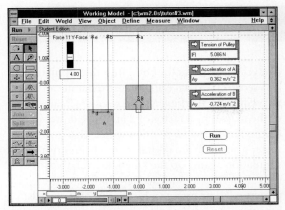

Figure 3.14

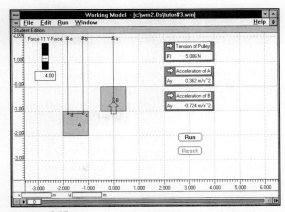

Figure 3.15

- Select **Edit** menu, **Player Mode**.

Your workspace has now changed considerably, as shown in Figure 3.15. The tool bar has disappeared and there are fewer menu choices. Note that there is a Run menu on the menu bar that you can use in Player mode. You will now run the simulation by clicking on the Run menu button that you added to the workspace.

- Click on the **Run** button.

The simulation starts to run.

- Click on the **Reset** button to stop and reset the simulation.

Player mode can be used for demonstrating Working Model simulations. Notice that the stop sign doesn't appear when the simulation runs, because you turned it off in the Preferences dialog box. You can switch back to Edit mode by selecting the Edit menu and Edit Mode.

- If you wish, **save** the simulation, and **exit** Working Model.

3.1 The Basics Set up the workspace with grid lines and rulers. Draw two circles (one with twice the radius of the other) and a pulley arrangement, as shown in Figure E3.1, but with the circles at the same initial height. Label the circles with appropriate names. Set up digital monitors for circle acceleration and pulley tension. Run the simulation and examine the meter's output to see if it makes sense. Add a downward force to the smaller circle, equal to its weight. Select the larger circle and define a control to adjust its mass. (You can control mass values just like you controlled force in this tutorial.) Run the simulation to examine the effects of the applied force and the variation of the larger circle's mass.

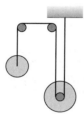

Figure E3.1

3.2 Applying the Basics You have just lost your engineering job and are trying to be an auto mechanic, but how do you get that engine out of the car so you can rebuild it? With the skills you have developed, create a Working Model simulation to represent the pulley arrangement of a "cherry picker," a device for picking engines out of vehicles. The large circle in Figure E3.2 represents the engine mass and the small circle is where a downward force is applied for lifting. Give the circles approximately the mass of an engine and a human being, respectively, and create acceleration

and velocity meters for both. Run the simulation and record the data. Can you use dynamic analysis to predict the Working Model results? Change one of the masses by a factor of 2 and repeat the simulation. Finally, add a position meter, change the meter's output from digital to graph, and run the simulation again. Do the kinematic variables (position, velocity, and acceleration) have the proper relationships graphically? In other words, how should position and velocity be related?

Figure E3.2

3.3 For Fun and Further Challenge Air-bearing floors are used by NASA to help simulate motion in space; as a NASA engineer, you need to simulationally determine the floor's friction characteristics. Create a block sliding on a surface (air-bearing floor) and connect the block to two circular masses with pulleys, as shown in Figure E3.3. Your idea is that the system motion is an indicator of the frictional force. Monitor the friction between the block and the surface by adjusting the friction properties of both identically, and adjust the block and circle masses to integer kilograms (1, 2, 3, …) to ease calculations. Predict the acceleration of the system as a function of the friction coefficient and verify it with the simulation. Change the friction and run the simulation again to confirm your theory. By the way, practically, what is wrong with this idea?

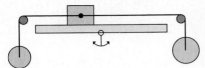

Figure E3.3

3.4 Design and Analysis   There is a new idea for an acceleration meter for a vehicle. The rectangle shown in Figure E3.4 is the meter casing; a roller is connected to the casing with a pulley. If the casing accelerates horizontally, the roller rolls sideways to a new stable position. Create constant casing acceleration by defining the box's velocity as **a*time** in the Properties window and anchoring the box, as you learned in Tutorial 2. Monitor the position of the roller and one of the pulley endpoints, and run the simulation several times to see the relationship between roller position and system acceleration. What do you think of this idea?

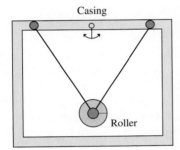

Figure E3.4

Exploring Your Homework   These problems deal with constant or varying forces acting on objects.

3.5   An airplane touches down on an aircraft carrier with a horizontal velocity of 50 m/s relative to the carrier. The arresting gear exerts a horizontal force of magnitude $10,000v$ N, where v is the plane's velocity in meters per second. The plane's mass is 6500 kg.

a. What maximum horizontal force does the arresting gear exert on the plane?

b. If other horizontal forces are neglected, what distance does the plane travel before coming to rest?

3.6   If a 15,000-lb. helicopter starts from rest and its rotor exerts a constant 20,000-lb. vertical force, how high does it rise in 2 s?

3.7   Two weights shown in Figure E3.5 are released from rest. How far does the 50-lb. weight fall in one-half second?

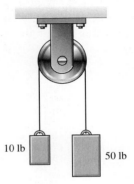

10 lb   50 lb

Figure E3.5

3.8   A 1000-slug rocket starts from rest and travels straight up. The total force exerted on it is $F = 100,000+10,000t-v^2$ lb. Using numerical integration with $\Delta t = 0.1$ s, determine the rocket's height and velocity for the first five time steps. (Assume that the change in the rocket's mass is negligible over this time interval.)

Exercises 3.5–3.8 are adapted from Bedford and Fowler's Engineering Mechanics: Dynamics *(Addison-Wesley, 1995), pages 104, 107, 110, 132.*

Satellite Trajectories

Objectives

In this tutorial you will learn to:
- Zoom on very large objects
- Define gravity
- Track
- Speed up a simulation
- Scale velocity vectors

Dynamics principles related to this tutorial

- Planetary motion
- Central force problems
- Conservation of angular momentum

Introduction

Many problems associated with space technology, such as planetary motion and satellite guidance problems, require us to think in very large proportions. In this tutorial, you will see that you can easily create Working Model simulations for very large-scale problems. You will learn how you can use different gravitational fields and how you can track motion, as well as other properties of objects, throughout a simulation. In addition, you will use Working Model to demonstrate the principle of conservation of angular momentum for an orbiting mass.

Problem Statement

An artificial satellite has been built to move among the planets and study their atmospheres. The satellite will first study the earth's atmosphere for a short period of time, and then proceed to the planet Mars to gather additional data. While it is orbiting the earth, it is initially at a height of $h = 7 \times 10^6$ m above the earth's surface, as shown in Figure 4.1. Figure 4.1 is not drawn to scale. Create a Working Model simulation that shows the satellite (1) in a circular orbit about the earth at height h, (2) in an elliptical orbit starting at height h, with a semi-major axis, $a = 20 \times 10^6$ m, and (3) starting from height h with a velocity that will cause it to escape from the earth's gravitational field (escape velocity). For cases (2) and (3), plot the paths of the satellite and show a velocity vector tracking with the satellite. The following quantities are given:

Earth radius = R_e = 6.372×10⁶ m (assume spherical and stationary)

Earth mass = M_e = 5.976×10²⁴ kg

Satellite radius = r_s = 100 m

Satellite mass = m_s = 1000 kg

Figure 4.1

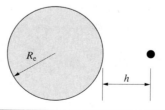

Creating the Components

Open Working Model by double-clicking on its icon. Resize your window if necessary before you begin.

- Click twice on the **Circle** icon on the tool bar to activate it.

If you need help, see Tutorial 1 for creating circles, and Tutorial 2 for creating multiple objects.

- Click twice in the workspace, creating two circles.
- Click on the **Arrow** icon on the tool bar to deactivate the Circle icon.

Your workspace should look like Figure 4.2.

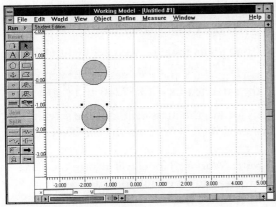

Figure 4.2

- Double-click on the first circle you made to open its **Properties** window.
- Change the values for **x** and **y** in the Properties window to **x = 0.0** and **y = 0.0**.

This locates the selected circle at the origin. This circle will represent the earth.

Tip Remember that the circle you made first will be called Mass[1]. You can determine which circle is Mass[1] by positioning your cursor over each circle and looking at the Help Ribbon at the top of the screen.

- Change the **mass** of the earth to **5.976e24** kg in the Properties window.

The format that Working Model will use for this number is 5.976e+24; your number will change to this default format when you press Enter or Tab, select a new text box in the Properties window, or leave the Properties window. Leave the window open for now.

- Choose **Window** menu, **Geometry...**

This opens the Geometry window. As you learned in Tutorial 2, you can use the Geometry window to size mass objects precisely. You can also use this window to relocate their center of mass (COM). You will see in Tutorials 11 and 12 how you can use the Geometry window to modify vertex locations for polygons and curved slots. In Tutorial 2, you changed the width and height of rectangles; here you will change the radius of the circle.

- Change the **radius** of the earth to **6.372e6** m in the Geometry window.

When you press Enter, note that the circle disappears except for its midpoint and a line, as shown in Figure 4.3. It is too large to display on your screen. Your second circle may or may not be visible, depending on where you placed it.

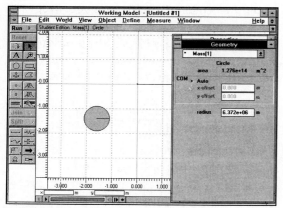

Figure 4.3

- Scroll the workspace so that the second circle is visible.

- Click on the second circle.

Now the Properties and Geometry windows show information for the second circle, Mass[2], in Figure 4.4. The mass numbers on your screen and in the figures should match if you selected Mass[1] to be the earth. This circle will be the satellite.

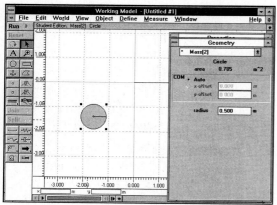

Figure 4.4

- Change the **radius** of the satellite to **100** m in the Geometry window.

This circle also disappears from the workspace, as shown in Figure 4.5.

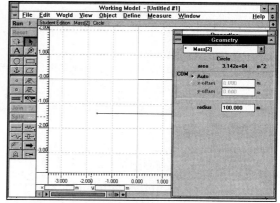

Figure 4.5

- Select the **Properties** window.

- Change the **mass** of the satellite to **1000** kg.

- Change the **x** and **y** values for the center of the satellite to **x = 13.372e6** and **y = 0.**

This locates the mass center of the satellite at a distance of $(R_e + h)$ from the earth's center.

> **Note:** Working Model will change the value 13.372e6 to its default format, 1.337e+07, when you enter the x location.

- Close the **Properties** and **Geometry** windows.

Your workspace should look like Figure 4.6. Only a line is visible on the screen, because the circle representing the earth is so large and the satellite is now outside the workspace.

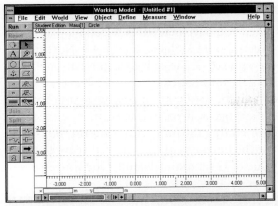

Figure 4.6

Zooming on Very Large Objects

In Tutorial 1 you learned how to zoom on objects. You have now created objects on a very large scale. Working Model is capable of handling such large objects; however, you must zoom the workspace out significantly in order to observe these objects.

- Select **View** menu, **View Size...**

The View Size dialog box shown in Figure 4.7 appears to allow you to select the scale at which you view your simulation. The upper box in the dialog box indicates the size of the workspace relative to objects in real life, i.e., the scale used for drawing objects. For example, a scale of 1 means that objects in the workspace are the same size as they are in reality. A scale of less than 1 means that objects are shown smaller than they are in reality, and vice versa. The Working Model default is to draw objects 0.021 times their

actual size. In this example, you want the earth to appear as approximately 15 mm (0.015 m) in size on the workspace, so you will use a scale of approximately $0.015/R_e = 0.015/6.372 \times 10^6 = 2.354 \times 10^{-9}$.

- Highlight the top text box, which appears in the line **Objects on screen are _____ times actual size**.

- Type **2.5e-09**.

- Press **Tab**.

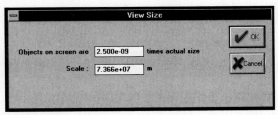

Figure 4.7

This makes the object on the screen 2.5×10^{-9} times its actual size. The lower text box, marked Scale (or Window Width on the Macintosh), indicates the scaled width of the workspace. You can change the value in either one of the boxes; the other value will then change automatically.

- Select **OK**.

- Click on the **scroll bars** to move the earth to the **right side** of the workspace.

- **Save** your simulation as **TUTOR#4**

Your workspace should look like Figure 4.8. The satellite is barely visible as a dot to the right of the earth, above the cursor in the figure.

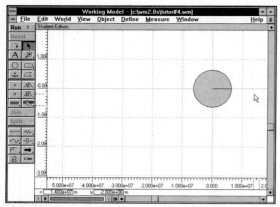

Figure 4.8

Defining Gravity

In Working Model there are three types of gravitational fields to choose from: none, vertical (the Working Model default), and planetary. If you choose one of the latter two options, you can alter the numeric value of the acceleration due to gravity.

- Select **World** menu, **Gravity...**

The Gravity dialog box appears on your screen. For vertical fields, you can select the acceleration due to gravity at the surface of the earth or moon, input any other value in the text box, or use the slider control. For this problem, you will select planetary gravitation. Planetary gravity simulates the true gravitational attraction that exists between all masses. When adjusting planetary gravity, you are changing the value of G, the universal gravitational constant. The forces exerted by gravity between objects, such as a person or a book, are minuscule because the objects are not massive enough. Thus, to see the effects of planetary gravity, your simulation must have one or more extremely massive bodies.

- Select **Planetary**.

When you have made this selection, your screen should look like Figure 4.9. Note that the value of G, the universal gravitational constant, is automatically set to 6.670e-11, the default for the Our Universe selection. You can change it, if you wish, by moving the slider or editing the numerical value in the box. For the purposes of this tutorial, leave the value of G set to the default, Our Universe.

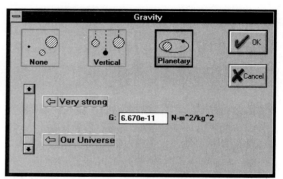

Figure 4.9

- Click **OK**.

Calculating the Circular Orbit

In order to proceed with developing the experiment, you need to determine the velocities for the satellite that will create the three desired conditions for the problem:

1. Circular orbit

2. Elliptical orbit

3. Escape from the earth's gravitational field

You will first treat the circular orbit. Using the formula,

$$v_c = R_e \sqrt{\frac{g}{R_e + h}}$$

where g = acceleration due to gravity at the earth's surface = 9.81 m/s^2, you can calculate that the velocity necessary for a circular orbit is 5458 m/s. You will create a velocity control to set this condition.

Using a Selection Rectangle

You need to select the satellite, but it is so small that you need to use a method other than clicking to select it. You will create a *selection rectangle* around the satellite.

- Click in the workspace above and to the left of the satellite and hold the mouse button down.

- Drag to the bottom right of the satellite.

As you drag, a dotted rectangle appears around the satellite, as shown in Figure 4.10. All other objects on the screen are deselected.

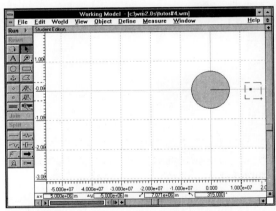

Figure 4.10

- Release the mouse button.

Tip You can also highlight hard-to-select objects by using the Properties window or by using the Select All option on the

Edit menu. To select with the Properties window, select any object on the screen and open the Properties window. You can then use the pull-down menu at the top of the window to select the object you want. To use Select All, choose the Edit menu, Select All, then hold the Shift key down and click on those objects you do not want in the highlighted selection. In this case, you would have selected both the earth and the satellite with Select All, and then deselected the earth.

Note that the satellite appears larger now; since you have selected it, it has the small corner blocks, which make it look larger. Refer to Figure 4.11.

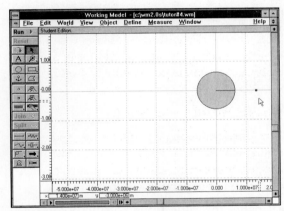

Figure 4.11

You will now create a control so that you can regulate the velocity of the satellite from the workspace. If necessary, refer to Tutorial 3, where defining controls was first introduced.

- Select **Define** menu, **New Control, Initial Y Velocity**.

A new initial Y velocity control appears in your workspace.

- Double-click on the title of the **Velocity** control to open its **Properties** window.

You will set the range on the values to be regulated by the control. You want the range to be $5000 < Vy < 8000$ m/s. You must set the maximum value first; if you set the minimum, 5000, first, it will automatically switch into the maximum box, since 5000 is larger than the default minimum.

- Set the **max** value to **8000** and the **min** value to **5000**.

Your workspace should look like Figure 4.12.

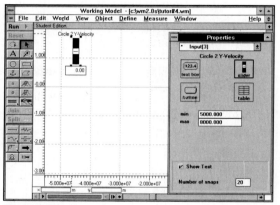

Figure 4.12

- Close the **Properties** window.
- Change the value of the **Y-velocity** control by typing **5458** in the text box below the slider control.

Tracking

It is sometimes instructive to have a "picture" of the path followed by an object in the course of a simulation. In this case, because the satellite is so small relative to the earth, such a picture would be very helpful. The Tracking feature leaves a trace on the screen of an object's movement at selected intervals. The Tracking submenu gives you a variety of options for setting the tracking interval, and an Other option that lets you enter any number of frames. You will select every frame. You will learn later in this tutorial why seeing less than every frame of an object's movement is sometimes desirable.

- Select **World** menu, **Tracking, Every frame**.

Running the Simulation

- **Run** the simulation.

The simulation runs very slowly. Let it run until it is clear that the path will be circular. Your workspace should look like Figure 4.13.

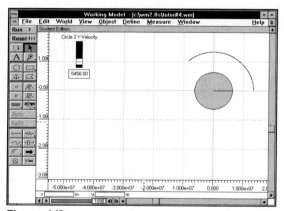

Figure 4.13

- Click on **Reset.**

Adjusting the Time Step

Every frame in a simulation represents a certain point in time. The time interval between frames is called the *time step,* and the accuracy of an experiment will be influenced by the size of the time step. In general, the larger the time step, the faster the experiment will run. Also, in general, the smaller the time step, the more accurate the experiment will be.

In most cases, you don't need to be concerned about the precise value of the time step. Working Model automatically calculates the value of the time step as it monitors your simulation. You can change this, however, by using the Simulation dialog box, shown in Figure 4.14.

■ Select **World** menu, **Accuracy...**

Figure 4.14

Working Model offers two preset options: Fast and Accurate. Fast uses a less rigorous method of integration and a larger time step. Accurate is the Working Model default setting. The third option, Custom, allows you to set the simulation variables yourself. You will learn more about accuracy and the time step in later tutorials; technical details are included in Appendix A.

The animation time step is the time interval between the frames of animation on your screen. It is not the same as the integration time step, which is the step used in the numerical integration methods used by Working Model. Working Model automatically defaults to an animation time step suitable for your simulation. You can override this

setting, however, and display fewer frames, by making the animation step larger. (This does not affect the integration time step, and a single animation step may contain several integration steps.)

■ Select the **lower radio button** in the Animation Step section of the dialog box.

■ Set the **animation step** to **150** s in the upper of the two text boxes.

■ Click **OK**.

The animation step is now set to 150 s per frame.

Obtaining Results

■ Click **Run**.

Since the animation step is much larger, the experiment now runs much faster. When your workspace looks like Figure 4.15,

■ Click **Stop**.

You have completed case (1) with your simulation of a circular orbit.

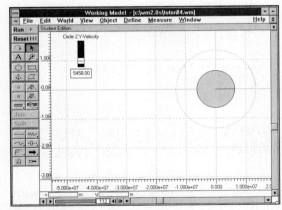

Figure 4.15

Creating the Elliptical Orbit

Now you will change the velocity of the satellite to achieve an elliptical orbit. The velocity necessary for the required elliptical orbit, whose starting point will be its closest point to earth, and whose major axis will be aligned with the present x axis, can be calculated with the formula,

$$v_e = R_e \sqrt{\left(\frac{g}{a}\right)\left(\frac{r_{max}}{r_{min}}\right)}$$

where $r_{min} = R_e + h$ = the initial distance of the satellite from the center of the earth, and $r_{max} = 2a - r_{min}$. This calculation results in a velocity of 6297 m/s, which you will test with your simulation.

- Change the value of the **y-velocity** control by typing **6297** in the text box below the slider.

- **Run** the simulation again through one complete earth orbit.

- Click **Stop.**

Your workspace should look like Figure 4.16.

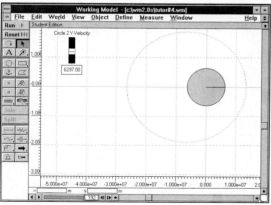

Figure 4.16

You will now plot the simulation with velocity vectors. First, using the Tracking feature, you will increase the number of frames between each plotted position. If a plot of the satellite's position and velocity were to be made at every frame, the vectors would crowd together.

- Click **Reset**.

- Select **World** menu, **Tracking, Every 8 Frames**.

- Select the **satellite,** using a **selection rectangle**.

- Select **Define** menu, **Vectors, Velocity**.

- **Run** the simulation through one complete earth orbit.

- Click **Stop**.

A small blue dot moves in the orbit, as shown in Figure 4.17; this is the velocity vector, but the default scale makes it too small to see. The velocity vector must be scaled to a larger size.

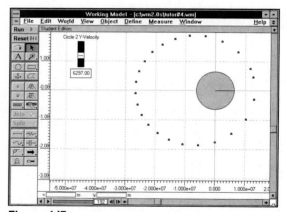

Figure 4.17

Scaling Velocity Vectors

You can adjust the lengths used to display velocity, acceleration, and force vectors by using the Vector Lengths dialog box. You will now change the length of the velocity vector so that you can see it in the workspace.

- Click **Reset**.

- Select **Define** menu, **Vector Lengths...**

The Vector Lengths dialog box appears, as shown in Figure 4.18. The velocity slider has a "short" value of 0.200. If you move the slider to the top with the cursor, you will see that its maximum (i.e., its "long" value) is 1.000. Although the slider limits are a minimum of 0.200 and a maximum of 1.000, you can enter a number below or above the range by typing it in the text box.

The number you enter in the Vector Lengths dialog box is multiplied by the measured value of the vector to determine the vector's scaled length. The resulting number is the unit length of the scaled vector in the current unit system.

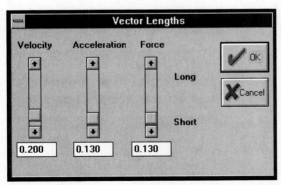

Figure 4.18

- Change the value in the text box below the **velocity** slider to **1000**.

- Click **OK**.

One large vector appears from the satellite, as shown in Figure 4.19. You will now erase the track that is already on your screen so that you can clearly view the new track.

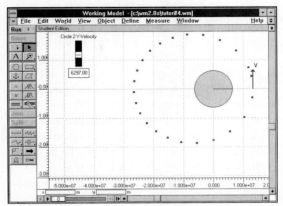

Figure 4.19

- Select **World** menu, **Erase Track**.

- **Run** the simulation again.

- Click **Reset** after the satellite has completed one orbit.

Your workspace should look like Figure 4.20. Case (2) of the Problem Statement is complete.

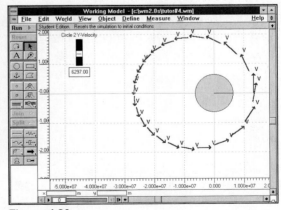

Figure 4.20

<table>
<tr><td>

Link to Dynamics

</td><td>

Planetary motion problems are examples of *central force problems*. The defining characteristic of a central

</td></tr>
</table>

force problem is that all the forces on a body (the satellite, in this case) pass through a central point (here, the mass center of the earth); thus, even though there is no conservation of linear momentum (i.e., the net force on the satellite is not equal to zero), there is conservation of angular momentum about the central point (i.e., the momentum of all forces about the central point is equal to zero). Figure 4.20 illustrates conservation of angular momentum about the earth's center. The magnitude of the satellite's angular momentum is $m_s v_t r$ (where v_t is the satellite's velocity tangent to the path of motion, and r is the distance from the earth's center to the satellite). Conservation of angular momentum requires that $m_s v_t r$ remains constant; you can see that as the satellite moves away from the earth (i.e., r increases), its velocity decreases, and vice versa.

Setting the Escape Velocity

You now need to determine the escape velocity for the satellite to satisfy the conditions for case (3).

The velocity necessary for the satellite to escape from the earth's gravitational field can be calculated with the formula,

$$v_{esc} = \sqrt{\frac{2GM_e}{R_e + h}}$$

You will change the velocity to 7721 m/s to test the calculations.

- Change the value of the **y-velocity** control by typing **7721** in the text box below the slider.

- **Run** the simulation until the satellite leaves the screen.

Your screen should look like Figure 4.21. The satellite has escaped from the earth's gravitational field, confirming the calculations and completing the problem.

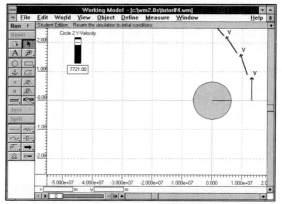

Figure 4.2l

- Click on **Reset**.

- **Save** your simulation and **exit** Working Model.

4.1 The Basics Set up Working Model with planetary gravity and experiment with the orbit of a 100-kg satellite around the earth. Let the satellite start at 2.e7 m from the earth and create an initial vertical velocity control for the satellite with a minimum of 0 and a maximum of 8000 m/s. Add a velocity vector for the satellite and a meter for time, and track the satellite's orbit. Run the simulation for different initial velocities. See which velocities cause the satellite to plummet to a fiery crash and which velocities result in a reasonable orbit. Monitor the period of successful orbits. You will certainly have to zoom out and adjust the animation step to see what is happening.

4.2 Applying the Basics On the fast track to become an astronaut, you are exploring the orbital mechanics of the moon. The eccentricity, e, of the moon's orbit about the earth is 0.055 and its mean distance to the earth, a, is 384,398 km. The mass of the moon relative to the earth is 0.0123, and its period of orbit is 27.32 days. With a Working Model simulation set up to look like Figure E4.1, determine the moon's velocity at perigee (minimum approach to earth) to get the orbit close to the correct period. Trial and error is a good tool to see the effects of velocity change. Calculate the correct velocity with help from your dynamics text. What is the velocity at apogee? You may want to anchor the earth to the background, use tracking, create a meter for time, etc. (Note: Eccentricity is a measure of how non-circular an orbit is. Minimum orbital radius is $r_{min} = (1-e)a$ and maximum orbital radius is $r_{max} = (1+e)a$.)

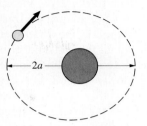

Figure E4.1

4.3 For Fun and Further Challenge The navigational computer on the Starship Enterprise is down, so manual (PC) calculations are necessary to determine which warp speed is appropriate to pass by an approaching binary star. In order to solve this problem, you need to predict the star's orbital motion. Assume that the two stars, one the mass of the sun and one double the mass of the sun, are orbiting each other as a binary star. (The sun is 330,000 times more massive than the earth.) Suppose they are separated by 50 diameters (although this is unrealistic), as shown in Figure E4.2. What happens if they have no initial velocity? What happens if the smaller star has twice the initial velocity of the larger star but in the opposite direction? (Make the velocities large enough to cause orbit.) What happens if they have arbitrary initial velocities? Use Working Model to examine these questions and use meters, vectors, and tracking to understand the dynamics of the process. What can you say about the linear and angular momentum about the center of the individual stars or the two stars together?

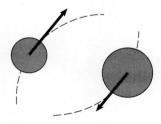

Figure E4.2

4.4 Design and Analysis Open or recreate
Exercise 2.4, the tire rolling down the road. Display
the velocity vector of the wheel, as shown in Figure
E4.3, and turn on tracking. Use the Appearance
window for the wheel to turn off Track outline, so
tracking will show only the wheel's center of mass.
Create a control for the initial horizontal velocity
of the wheel and run the simulation. Adjust the
scale on the velocity vector if necessary. Experiment
with tracking less than every frame, changing the
animation step of the simulation under Accuracy,
using different initial velocities, and running under
the influence of lunar gravity. Report on your
results.

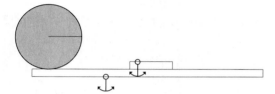

Figure E4.3

Exploring Your Homework These exercises
involve orbital mechanics and central force problems. Do you need to know the satellite's mass to
do these problems? Try them with several values of
mass, from 10 to 1000 kg.

4.5 A satellite at $r_0 = 10,000$ miles from the center of the earth is given an initial velocity,
$v_0 = 20,000$ ft/sec, in the direction shown in Figure
E4.4. Determine the magnitude of its transverse
component of velocity when $r_0 = 20,000$ miles. The
radius of the earth is 3960 miles.

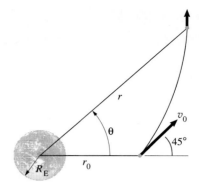

Figure E4.4

4.6 When an earth satellite is at perigee (the
point at which it is nearest to the earth), the magnitude of its velocity is $v_P = 7000$ m/s and its distance from the center of the earth is $r_P = 10,000$
km, as shown in Figure E4.5. What are the magnitude of its velocity v_A and its distance r_A from the
earth at apogee (the point at which it is farthest
from the earth)? The radius of the earth is
$R_E = 6370$ km.

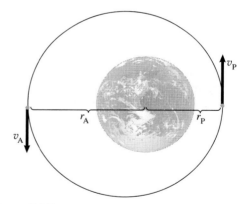

Figure E4.5

*Exercises 4.5 and 4.6 are adapted from Bedford and
Fowler's* Engineering Mechanics: Dynamics
(Addison-Wesley, 1995), pages 212, 214.

Rolling Contact

Objectives

In this tutorial you will learn to:
- Define the angular velocity of an object
- Create actuators and connect them to objects
- Create and locate measurement points
- Determine the names of objects in the Properties window
- Track selected objects
- Obtain graphs from meters
- Use the Start Here command

Dynamics principles related to this tutorial

- Rolling contact
- Instant centers
- Epicycloids

Introduction

In this tutorial you will develop a model of rolling contact between two friction wheels. In addition to friction wheels, rolling-contact kinematics describe the relative motion in other important types of machines, such as friction belts and gear trains. You will connect two objects by means of a Working Model actuator, and you will gain an understanding of the way that Working Model assigns names and numbers to the objects and constraints that you create. In addition, your simulation will demonstrate the instantaneous center of velocity during rolling contact.

Problem Statement

The system shown in Figure 5.1 represents a toy spirograph; a child places the spirograph on a piece of paper, places a colored pen at different positions along line AB, and traces the motion on the paper. Very attractive patterns emerge (curves called *epicycloids*). Create a Working Model simulation that represents this system. The inner wheel (radius r_1 = 1 m) is fixed. A rod is pinned to the centers of both wheels and causes the outer wheel (radius $r_2 = 0.5$ m) to roll around the inner wheel. The outer wheel's angular velocity is π rad/sec. Show plots of the velocity of points at the center and outer rim of the smaller wheel as functions of time. Assume that the outer wheel rolls without sliding and that the gravitational effects are negligible. This problem exemplifies how epicyclic or planetary gear trains behave. You will learn more about epicyclic gear trains in Tutorial 13.

Figure 5.1

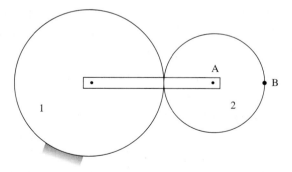

Creating the Components

Open a new Working Model document. Resize your window as necessary before continuing.

- Create two circles on the x axis with their mass centers at **x = 0** and **x = 1.5**, respectively.

Tip You can use the Coordinates displays above the Tape Player controls at the bottom of your workspace to help you place objects accurately in the simulation.

Your workspace should look like Figure 5.2.

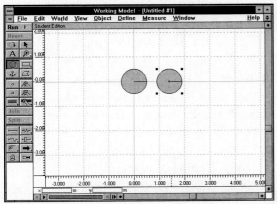

Figure 5.2

- Make sure the **Arrow** tool is active before you continue.
- Use the **Geometry** window to change the **radius** of the left-hand circle to 1 m.
- Leave the **radius** of the right-hand circle at its Working Model default, **0.5 m**.
- Close the **Geometry** window when you are finished.

The surfaces of the two circles are now in contact with one another. Your screen should look like Figure 5.3. If it doesn't, use the Properties window to check the x and y position of the two circles.

Tip You can use the scroll bar to move the objects in the workspace so that you can view both objects, even when multiple windows are open. If necessary, move the window by clicking on and dragging its top border to get to the scroll bar.

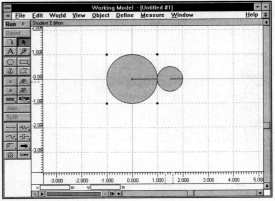

Figure 5.3

Defining Angular Velocity

The angular velocity of the smaller circle is given as π rad/sec. You will input this value in the smaller circle's Properties window, in the box labeled Vø. The units shown indicate °/s (degrees/sec); π rad/sec = 180 °/s.

- Use the **Properties** window for the small circle to change the value for **Vø to 180** °/s.
- Close the **Properties** window when done.

Since the problem stated that gravity would be neglected, you should now turn Gravity off.

- Select **World** menu, **Gravity...**

- Select **None**.

- Click **OK**.

The larger circle requires an anchor because the problem stated that it is fixed in space.

- **Anchor** the larger circle to the background.

If you need help using anchors as positional constraints, refer to Tutorial 2. Your workspace should look like Figure 5.4.

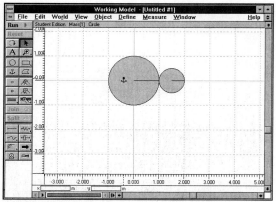

Figure 5.4

Creating Actuators

Actuators are multi-purpose objects that can exert the forces necessary to constrain lengths, velocities, and accelerations. There are four types of actuators:

1. *Force* actuators exert a specified force between their two endpoints.

2. *Length* actuators exert whatever force is necessary to keep their endpoints a particular distance apart.

3. *Velocity* actuators exert whatever force is necessary to keep their endpoints moving together or apart at a specific velocity.

4. *Acceleration* actuators exert whatever force is necessary to keep their endpoints moving together or apart at a specific acceleration.

An actuator is necessary in this problem to model the rod connecting the wheel centers, in order to maintain steady contact between the two circles. Although Working Model does have a Rod tool available, a Rod in this model would not maintain a constant contact force between the circles, but would only keep the circles a particular distance apart. This might cause the small circle to rotate in place. A force actuator imposes a constraint that will keep the contact force uniform, causing the small circle to roll on the large circle.

- Click on the **Actuator** icon on the tool bar (the icon at the bottom right corner of the tool bar).

- Click on the center of the large wheel and drag from the center of the large wheel to the center of the small wheel.

An actuator is created; its endpoints should be the center of the large wheel and the center of the small wheel. Your screen should look like Figure 5.5.

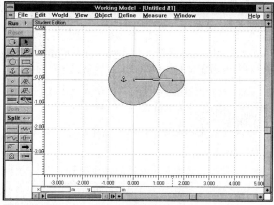

Figure 5.5

At this point, you can select the type of actuator you want to use by opening the actuator's Properties window, as shown in Figure 5.6.

- Open the actuator's **Properties** window.

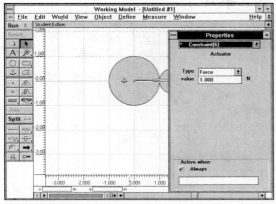

Figure 5.6

You can choose one of four kinds of actuators from the Type pull-down menu: Force, Length, Velocity, or Acceleration. In this case, you will use the default Force actuator because you want to make sure that a constant contact force is always maintained between the two wheels. The default value of the force is a positive 1 N. A positive value means that the force will tend to increase the distance between the endpoints of the actuator. In this problem you want the actuator to help maintain contact, or to tend to make the distance between its endpoints smaller; a negative value will achieve this. Therefore, you will make the value of the actuator force –1 N.

- Change the **force** value to **–1** N.

You can give actuators more complex behaviors by entering a formula in the value box of the Properties window; you can also control when they are active by entering a value or formula in the Active when box at the bottom of the window. You saw how to use formulas briefly in Tutorial 2, and you will see more advanced formulas in Tutorial 10. (Terminology for all Working Model formulas is provided in Appendix B.) The Active when option will be discussed in Tutorial 8.

Creating Measurement Points

You can create stand-alone point elements to measure properties at a specific location on a mass. You will attach a point to the rim of the smaller wheel to compare the velocity and acceleration at that location to those at its center. You will also be able to track the path of this point, as a spirograph would.

- Scroll the workspace so you can see both circles (if necessary).

- Click on the **Point element** icon on the tool bar (below the anchor).

- Click on the smaller wheel at a point just inside the outer rim.

  Tip If you click too close to the edge, the point may be attached to the background reference frame instead of to the wheel. It is better to be certain that it is attached to the wheel first, and then move it to the rim.

You will use the point's Properties window to move the point to the rim of the wheel.

- Change the value of **x** in the point's Properties window to **0.5.**

- Change the value of **y** to **0** (if necessary).

- Press **Enter** to accept the coordinate changes while keeping the Properties window open and active.

Once a point is attached to an object, Working Model measures its location ((0.5,0.0) in this case) with respect to the mass center of the object. The global coordinates at the bottom of the Properties window show the location of the point with respect to the workspace x and y axes. Now the point is located on the rim, as shown in Figure 5.7.

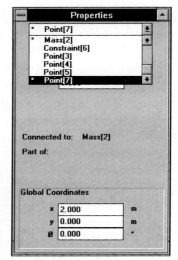

Figure 5.8

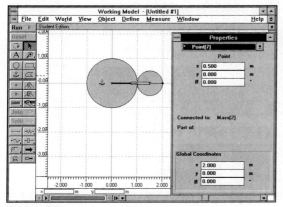

Figure 5.7

Determining Object Names

To work this problem, you need a point at the mass center of the smaller wheel in order to measure its velocity. This point already exists. It was created when you defined the actuator. You will examine the Properties window a little more closely to determine which point it is.

W Open the pull-down menu at the top of the **Properties** window by clicking once on the **down arrow**.

W Your screen should look like Figure 5.8.

M Open the pull-down menu at the top of the **Properties** window by clicking on the **down arrow** and holding the mouse button down.

M Your screen should look like Figure 5.9. (You may have to move the mouse up a bit while still holding the mouse button down to reveal more of the list.)

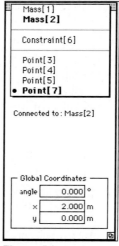

Figure 5.9

Note that each item has a number attached to it. As you may recall from Tutorial 1, these numbers are assigned in the order in which you create the objects. If you created your circles as instructed, your numbers should match those in Figure 5.8 or Figure 5.9 and in the table below. (Point[4] and Point[5] are reversed on the Macintosh; substitute the appropriate number in the discussion below.) If your numbers are different, make a note of them on the table. You will look at them in detail.

The object now highlighted Select	is the:
Mass[1]	Large wheel
Mass[2]	Small wheel
Point[3]	Anchor
Point[4]	Mass center of small wheel
Point[5]	Mass center of large wheel
Constraint[6]	Actuator
Point[7]	Point on rim of small wheel

When you create meters, they will also be assigned numbers; e.g., the next object you create will be a meter, and will appear in the Properties window pull-down list as Output[8]. The two points whose velocity you want to measure are Point[7], the rim point, and Point[4], the center point of the small wheel. You also want to trace the path of Point[7], the rim point.

- Click on **Constraint[6]** on the pull-down menu.

- Click on the **down arrow** next to the pull-down menu.

Note that three of the items on the pull-down menu in Figure 5.10 have the symbol * in

front of them (in Windows; on the Macintosh they are in bold): Constraint[6], Point[4] (the center of the small wheel), and Point[5] (the center of the large wheel). This tells you that the actuator is connected to Point[4] and Point[5].

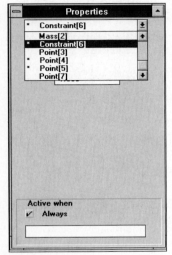

Figure 5.10

- Click on **Mass[1]** on the pull-down menu.

- Click again on the **down arrow** of the pull-down menu.

Note that now three different items have the symbol * in front of them (or are bold): Mass[1], Point[3], and Point[5]. This tells you that the large wheel, Mass[1], is connected to Point[3] (the anchor) and Point[5] (the center of the wheel and endpoint of the actuator). (On the Macintosh, they are in boldface and Point[4] is the center of the wheel and endpoint of the actuator.) You can always use this feature of the Properties window to clearly identify the connectivities of various objects and constraints.

Tracking Selected Objects

In Tutorial 4 you used tracking to trace the motion of all objects in the workspace (in that case, only the satellite was moving). You can also use tracking to trace the motion of selected objects or points. To do this, you must first make sure that tracking is turned off for all objects, and then turn it on for the object(s) of interest. To turn off all tracking:

- Close the **Properties** window.

- Select **Edit** menu, **Select All**.

This option selects every object that has been created. The small black corner squares that indicate that an item has been selected are displayed for all items, as shown in Figure 5.11.

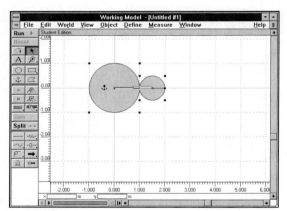

Figure 5.II

- Select **Window** menu, **Appearance...**

This is an easy way to change the appearance of several objects at once. There are three items in this window that determine the nature of the trace left on screen by an object when tracking is turned on:

1. The Track center of mass button leaves a point at each location of the mass object's center.

2. The Track connect button draws a connecting line between each location of the mass object's center of mass.

3. The Track outline button leaves a trace of the mass object's outline at each location.

You will set all of these to their off positions, as shown in Figure 5.12.

Figure 5.12

- Click **Track center of mass** until it is not checked.

- Click **Track outline** until it is not checked.

- Click in the workspace away from any object to deselect all objects.

The small black corner squares disappear.

- Click on **Point[7],** on the rim of the small wheel.

- Click on **Track connect** in the **Appearance** window so it is checked.

A line will trace the location of the point when you run the simulation. (Track center of mass and Track outline don't appear here because a point has no mass or outline.) If you selected Track, it would leave an image of the point on the screen at each frame, as you saw in Tutorial 4. Because the tracking for the small wheel is turned off, the track of the selected point will not be obscured by the track of the wheel.

Measuring Velocity

You will need meters to show the velocity for the points at the center and the rim of the small wheel.

- Create **velocity** meters with a **graph** display for the center (for Windows, **Point[4];** for the Macintosh, **Point[5]**) and the rim of the small wheel (**Point[7]**), respectively.

- Move the meters to locations where they will be easy to read.

Notice that your meters are labeled Output[8] and Output[9] in the Help Ribbon as you hold your cursor over each meter. Your workspace should look like Figure 5.13.

Use the Appearance window to label the elements of your simulation.

- Label the center point of the small wheel **A**.

- Label the rim point of the small wheel **B**.

- Name the meters so that they clearly identify the information they will display.

- Eliminate all but |**V**| on the output meters.

Refer to Tutorial 3 if you need help using the Appearance window. Your workspace should look like Figure 5.14.

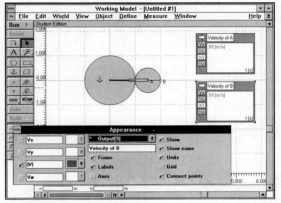

Figure 5.14

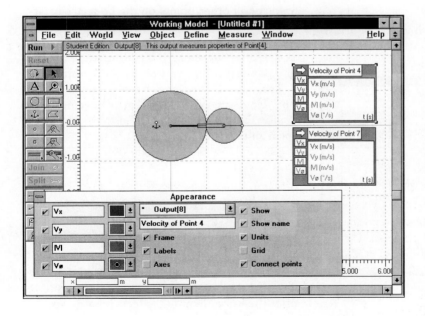

Figure 5.13

Since you created graphs to display the output data, it is desirable to see axes and grid lines on those graphs. You can turn on grids and axes for the graph display meters in the Appearance window.

- Click on the **Axes** and **Grid** buttons for both meters in the Appearance window.

- Close the **Appearance** window.

Tip You can use the Shift key to select both meters and change their Appearance windows at the same time.

The meters now have axes and grid lines.

- **Save** your simulation as **Tutor#5**

Running the Simulation

- **Run** the simulation until the small wheel has made 1.25 complete turns around the large wheel.

Your workspace should look like Figure 5.15 at the end of 1.25 turns.

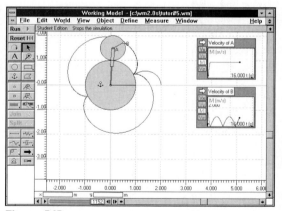

Figure 5.15

- Click **Reset**.

You can see that the magnitude of the velocity of the mass center of the small wheel, point A, is constant throughout the motion (except at the very start of the motion, because the system starts from rest). The rim point has a varying speed, which becomes zero when it comes into contact with the large wheel. There are some initial dynamic effects when the velocity of point A goes from zero to its constant value. These effects disappear after the first cycle; that is why the first portion of the tracking curve is not repeated. (You can eliminate them from your output by using the Start Here command on the World menu, as explained below).

Link to Dynamics

When point B is the contact point, it is known as the *instant center of velocity*. As long as there is no slip between the contacting surfaces, and one of the objects is fixed, the contact point will have zero velocity; all other points on the body are rotating about the contact point at this instant. Because point A has a constant speed, its tangential acceleration will be zero throughout the motion, and its normal acceleration will be constant.

The curve that point B traces is called an epicycloid, and two gears in an arrangement similar to the friction wheels in this problem form an *epicyclic gear train*.

Using the Start Here Command

You can use the current condition of a simulation as the initial condition for a new simulation. You will arrange to start your simulation after the initial dynamic effects have disappeared, i.e., after point B has made initial contact with the large wheel.

> **Tip** If you think you might ever need the old initial conditions, be sure to save the simulation before you use the Start Here option.

- Drag the frame indicator of the **Tape Player** to frame **128**.
- Select **World** menu, **Start Here**.

The current frame becomes frame 0 and the track disappears. The new initial conditions are set. The original initial conditions are lost, and the simulation is recalculated from the new start point.

- **Run** the simulation through 1.25 turns.
- Click **Reset**.

No transient effects appear in Figure 5.16, because the simulation now runs smoothly from the beginning.

Modifying the Simulation

Try moving the tracking point to other locations along line AB and change the color of the track in the Appearance window to get the real effect of a spirograph. What do the new epicycloids look like?

- **Save** your simulation and **exit** Working Model.

Figure 5.16

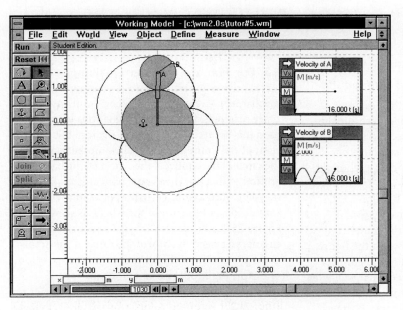

5.1 The Basics Create a 1-m square at the origin and place a circle with a 0.5-m radius next to it, just touching the square. Connect an actuator with 2 N of compressive force between their centers, as shown in Figure E5.1. Position points A and B as in the figure, track point B, and run the simulation for various initial angular velocities for the circle. Try 200 °/sec and 500 °/sec. See if you can understand why the system behaves differently for these two starting conditions. A meter for angular velocity may help.

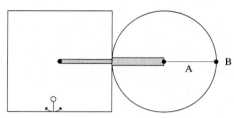

Figure E5.1

5.2 Applying the Basics In the design of warehouse automation, you need to examine the dynamics of a lift system. With a scaled model, you want to determine how powerful an activator needs to be to lift inventory. Create an anchored rectangle to represent the ground and an anchored rectangle at its left end, as shown in Figure E5.2. Create two 1-kg small rectangular masses and place one on the ground close to the center, and one off the end and down, as shown in the figure. Connect the two masses by a pulley and connect the mass resting on the ground to the left wall with an actuator. Select a length actuator with the prescribed length of 0.25*cos(4*t); where t is time. With force vectors and meters, examine the actuator and pulley tensions during the simulation. Can you calculate what these forces would be with your knowledge of free body diagrams and dynamics?

Change the actuator to a force actuator of 1*cos(4*t) and examine the acceleration of the vertical mass. How does this compare with your calculations? For a little fun, change the amplitude of the actuator force to 3 and let it rip!

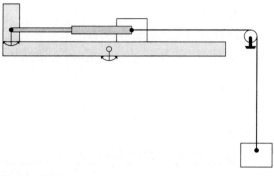

Figure E5.2

5.3 For Fun and Further Challenge As a locomotive engineer, you are examining a power mechanism. Create a circle (wheel) at the workspace origin and use the pin joint (you haven't used this before; it's located on the tool bar to the right of the Point icon) and click on the center of the circle. This connects the wheel to the background with a pin (bearing) so it can spin but not translate. Add a force actuator between the horizontal axis and the top of the wheel, as shown in Figure E5.3. In order to "invent" a reciprocating engine, you need to control the force of the actuator as a function of the wheel's rotational position. The Working Model variable formula for the wheel's rotational position is mass[1].p.r. (Refer to Tutorial 2 if you need to review variable formulas.) Design a force for the actuator that is a function of mass[1].p.r and see if you can get this train rolling. Use meters with graphical display to examine forces and velocities.

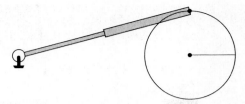

Figure E5.3

5.4 Design and Analysis Build a suspension test stand by anchoring 2 parallel walls 0.5 m apart. Add a 200-kg top mass (sprung mass) and a 20-kg bottom mass (unsprung mass) that fit perfectly between the walls. Set all friction values to zero so the sides do not interfere with the motion. Connect the bottom mass to the ground with a length actuator and connect the 2 masses together with a spring (which you haven't used yet; it's located in the right-hand column of the tool bar, below the Split icon). Attaching a spring is just like attaching an actuator. Click on the Spring tool and then click and drag from the center of the sprung mass to the center of the unsprung mass, as shown in Figure E5.4. Use the Properties window to set the spring constant, K, to 20,000. Set the actuator length to a sine wave with an amplitude of 0.25 m, using the formula 0.25*sin(*freq**t), where t is time. Run the experiment with 5 frequencies, *freq*, between 1 and 10 rad/sec. Use meters to monitor the motion of the masses. Do certain frequencies influence the system uniquely?

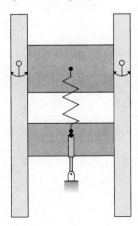

Figure E5.4

Exploring Your Homework The following problems look at tethered masses rotating about a central point. The tether can be modeled as a force actuator where the force is a function of other system variables.

5.5 A 2-kg disk slides on a smooth horizontal table and is connected to an elastic cord whose tension is $T = 6r$ N, where r is the radial position of the disk in meters. If the disk is at $r = 1$ m and is given an initial velocity of 4 m/s in the transverse direction, what are the magnitudes of the radial and transverse components of its velocity when $r = 2$ m?

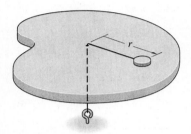

Figure E5.5

5.6 In Exercise 5.5, determine the maximum (or minimum) value of r reached by the disk.

Exercises 5.5 and 5.6 are adapted from Bedford and Fowler's Engineering Mechanics: Dynamics *(Addison-Wesley, 1995), page 213.*

Two Double-Pendulum Systems

Objectives

In this tutorial you will learn to:
- Use Show Center of Mass
- Use ropes
- Use Lock Points and Lock Controls
- Use the Do Not Collide option
- Switch between simulations
- Split and join objects
- Work with the scale of graphs

Dynamics principles related to this tutorial

- Newton's laws for rigid bodies
- Rotational inertia

Introduction

The two simulations described in this tutorial demonstrate the effects of rotational inertia in rigid-body motion. This has important implications for the design of many engineering devices, including gear trains, turbines, and drive mechanisms. In this tutorial, you will create two Working Model simulations and learn how to switch between them. You will learn how to use Working Model ropes to connect bodies, and you will improve some of your skills in building experiments and displaying output results. You will also demonstrate the importance of rotational inertia in rigid-body motion.

Problem Statement

Two double-pendulum systems are shown in Figure 6.1. The pendulums swing past each other without colliding. Both systems contain identical sliders on frictionless surfaces with two pendulums symmetrically attached to the centers of the sliders. The dimensions of the two identical sliders are 1 m X 0.5 m, and the radius of all the circles is 0.5 m. The only difference between the two systems is that in the system on the right, one of the pendulum ropes is attached at the top of the circle and the other is attached to the mass center of the circle. In the system on the left, both pendulum ropes are attached to the mass centers of the circles. Create separate Working Model simulations for each system, and demonstrate how to easily switch from one simulation to the other. Show the tensions in all ropes, the linear accelerations of the sliders, and the angular accelerations of all circles, then compare the two systems.

Figure 6.1

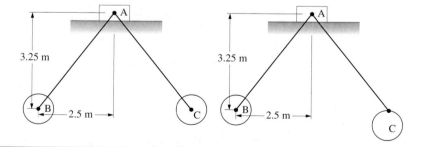

Setting Up Your Workspace

- Open a new Working Model simulation and resize your workspace window if necessary.

You will first change the numerical integration settings to Fast, so your simulation will run faster. You will experiment further with the Accuracy dialog box later in this tutorial.

- Select **World** menu, **Accuracy...**
- Select **Fast**.
- Click **OK**.

Creating the Components

First you will create a rectangle that is 7 m × 1 m to serve as the frictionless surface.

- Double-click on the **Rectangle** icon.
- Click and hold at the coordinates **x = -3.5 m, y = -.5 m**.
- Drag to the coordinates Δ**x** = **7 m**, Δ**y** = **1 m** and release the mouse button.

   Tip Remember that the Coordinates display at the bottom of your screen will change to Δx and Δy once you start to create your rectangle.

- Click and hold at the coordinates **x = -.5 m, y = .5 m**.
- Drag to the coordinates Δ**x** = **1 m**, Δ**y** = **.5 m** and release the mouse button.

According to the Problem Statement, your circles will have the Working Model default radius of .5 m, so you can simply click in the workspace to create them. You will learn to create circles by dragging in a later tutorial.

- Double-click on the **Circle** icon.

- Click at the coordinates **x = -2.5** m, **y = 1.5** m.
- Click at the coordinates **x = 2.5** m, **y = 1.5** m.
- Click the **Arrow** tool.

Your workspace should look like Figure 6.2. If it does not, drag the objects with the mouse or use the Properties window to reposition them.

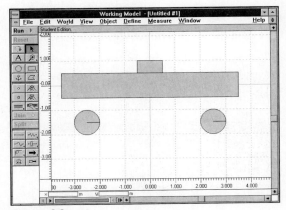

Figure 6.2

Because the larger rectangle represents the ground, you will place a positional constraint on it to anchor it to the workspace.

- **Anchor** the large rectangle.

The Problem Statement defines all surfaces as frictionless. You will now turn friction off for all objects.

- **Select all** objects (**Edit** menu).
- Open the **Properties** window.
- Set **stat.fric** to **0**.
- Set **kin.fric** to **0**.
- Close the **Properties** window.

Selecting Show center of mass

The center of each mass can be shown or hidden in Working Model. The default setting is to hide them. To make placement of the ropes easier, you will first display the center of mass for each object.

- If you have clicked anyplace in the workspace to deselect the objects, reselect **all** objects.

- Select **Window** menu, **Appearance...**

- Click **Show center of mass** so it is checked.

- Close the **Appearance** window.

The symbol X now marks the center of mass of each object. Your screen should now look like Figure 6.3.

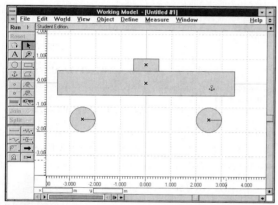

Figure 6.3

Using Ropes

Ropes prevent objects from separating by more than a specified distance. They will go slack (and have no effect) when the objects they are connected to move closer together than their length. You can attach ropes between one mass object and the background, or between two mass objects. You will now create the pendulum ropes for your experiment.

- Click on the **Rope** icon on the tool bar (the

third icon from the bottom in the left column).

- Click and drag from the center of mass of the left-hand circle to the center of mass of the small rectangle.

A rope appears, as shown in Figure 6.4.

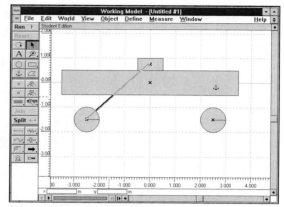

Figure 6.4

- In the same manner, create a **Rope** connecting the center of mass of the right-hand circle with the center of mass of the small rectangle.

Your workspace should look like Figure 6.5.

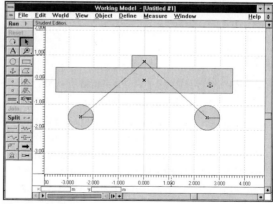

Figure 6.5

- Use the **Properties** window to identify the points that make up the endpoints of the ropes.

If you need help identifying connected points using the Properties window, refer to Tutorial 5.

Constraint[8] is the first rope. Figure 6.6 shows that its endpoints are Point[6] and Point[7]. Depending on the order in which you created your simulation, these numbers may not match exactly.

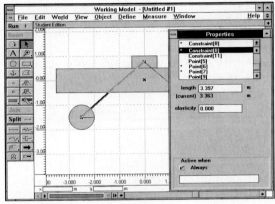

Figure 6.6

Make sure that all of the endpoints are located at (0,0) on their respective masses. (Remember that x and y coordinates for a point are relative to the center of mass of the object to which it is attached.) Choose Point[6] from the pull-down menu in the Properties window. Its Properties window shows that the (x,y) values for this point are (0,0), as shown in Figure 6.7. If the coordinates for your Point[6] are not (0,0), change them now. Check the other endpoint of Constraint[8] and the endpoints of the other rope as well.

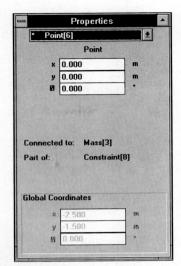

Figure 6.7

You no longer need Show center of mass.

- Use **Select All** and the **Appearance** window to turn off **Show center of mass**.

- Close the **Appearance** and **Properties** windows.

- Click anywhere in the workspace to deselect the objects.

Using Lock Points

When you select Lock Points, it becomes impossible to drag any point entities that you have created. This includes endpoints of constraints, such as the points at the ends of the ropes created for this simulation. This prevents you from accidentally grabbing and moving a previously positioned endpoint while trying to perform some other operation. Points that are locked cannot move with respect to the masses to which they are attached; if the masses move, so do the points. Lock Points does not prevent intentional relocation of points using the Properties window, however; you will use this function later in this tutorial.

- Select **View** menu, **Lock Points**.

- **Save** your simulation now as **Tutor#6A**

Adjusting the Time Step

- **Run** your simulation until the two circles collide.

Your circles should collide at about Frame 43, as shown in Figure 6.8. They will move away from each other after colliding.

Stop the simulation and use the Forward and Backward Step to move to Frame 43. (Frame numbers may vary a little on your system.) Notice that your circles overlap. In Frame 44 they are still overlapping, but they are beginning to move away from one another again. You can eliminate the overlap by changing the Accuracy settings once again.

- **Reset** your simulation.

- Select **World** menu, **Accuracy...**

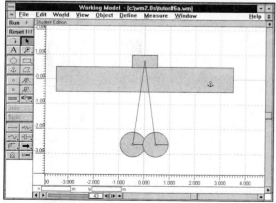

Figure 6.8

W Select **More** at the **right-hand side** of the dialog box to see more options.

M Select **More Choices** at the **left-hand side** of the dialog box to see more options.

The dialog box expands to reveal more options for controlling the time step and integration method used by Working Model, as shown in Figure 6.9.

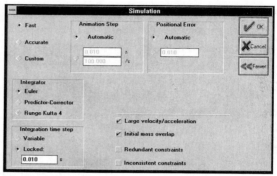

Figure 6.9

In Tutorial 4, you learned that the time step used for numerical integration is not always the same as the animation time step. You can control each separately in the Simulation dialog box. You learned how to change the animation step in Tutorial 4; the Integration time step may be set to Variable or locked when the dialog box is expanded.

- Click **Variable** for the **Integration time step.**

- Select **OK.**

- **Run** your simulation again.

Notice that the circles again collide at Frame 43, but they simply move away from one another after that; they do not overlap.

When accuracy is set to Fast, Working Model uses a Locked, or constant, time step. Notice that the Locked setting for the integration time step is paired with an input box. When the time step is locked, changing the value in this box changes the size of the integration time step, but the step is not allowed to vary during the course of the simulation. When the integration time step is allowed to vary, Working Model adjusts it to accommodate high acceleration or velocities by calculating more steps for a more accurate simulation. In order to eliminate the overlapping of the circles when they collide, you must set the integration time step to variable so Working Model can adjust the time step at the point of collision.

Remember that you started making changes to the Accuracy settings from the defaults set by Fast. Try changing to the Accurate mode and back to Fast and notice how the default settings change. (The integration step is variable by default in Accurate mode.) When locked, the integration step can be set independently from the animation step, making it possible to select a very small integration step for high accuracy, but to select a large animation step allows the results to display less frequently. The animation step must always be larger than the integration step, however.

- Reset your **Accuracy** settings to **Fast** before you continue.

Using the Do Not Collide Option

As you have seen, Working Model mass objects that are not connected to each other will collide when they come into contact. You will now prevent the pendulums from colliding so they can swing freely past each other.

- Click on one of the circles.

- Hold **Shift** down; click on the other circle to select it too.

Tip Remember, you can select and deselect any number of objects while holding Shift down.

- Select **Object** menu, **Do Not Collide**.

- **Run** the simulation until the bodies pass each other without colliding.

This time your circles should pass each other by about Frame 53.

- **Reset** your simulation.

- Use the **Appearance** window to label the **small rectangle**, which is the slider, **A**; the **left-hand circle**, **B**; and the **right-hand circle**, **C**.

- Select **Show name** so the labels appear.

- Close the **Appearance** window when done.

Your workspace should look like Figure 6.10.

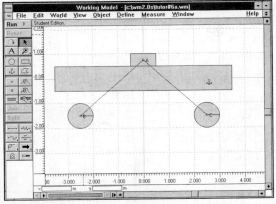

Figure 6.10

Creating Output Meters

Create five output meters displaying the parameters listed in the table below and label them as indicated. Different meters will be created with different defaults (some digital and some graphical). Set the display mode of each meter as specified in the table. If you need help creating meters, refer to Tutorial 1.

W Remember that the order in which the displays change is: digital → graph → bar.

Your workspace should look like Figure 6.11.

You have finished modeling the first system in the problem. The second pendulum system is very similar to the first; therefore, you will use the current simulation to set up the second.

- Save **TUTOR#6A** again.
- Select **File** menu, **Save As...**
- Name this same file **TUTOR#6B**

W Select **OK.**

M Select **Save.**

You have created two identical files: TUTOR#6A and TUTOR#6B. Note that you are now working in file TUTOR#6B.

Creating Menu Buttons

As you learned in Tutorial 3, a menu button enables you to add frequently used commands directly to the workspace. You can create menu buttons to perform any menu task. Clicking on a menu button is exactly the same as selecting the representative command from its menu. The purpose of the menu button you will create here is to switch between simulations.

- Select **Define** menu, **New Menu Button...**

W The New Menu Button dialog box opens with the available choices of menu items you can choose for this button.

W Select **Open** from the choices in the dialog box.

W Click **OK.**

M The New Menu Button dialog box opens and asks you to select the operation you want the new button to perform. You should select the operation from the Working Model menus.

	Measure...	...With Meter titled...	...As a...
1	Tension in left-hand rope	Tension Rope AB	Digital display
2	Tension in right-hand rope	Tension Rope AC	Digital display
3	X-Acceleration of slider	X Accel. A	Graphical display
4	Rotational acceleration of left-hand circle	Ang. Accel. B	Bar graph display
5	Rotational acceleration of right-hand circle	Ang. Accel. C	Bar graph display

Figure 6.11

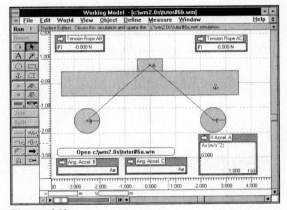

Figure 6.11 *(upper figure)*

M̄ Select **File** menu, **Open...**

The Open File dialog box appears, with a list of files.

- Select the file **TUTOR#6A**.

W̄ Click **OK**.

M̄ Click **Open,** then **OK**.

W̄ Remember that Working Model adds the file extension .WM to your file when it is saved, so you are really looking for TUTOR#6A.WM.

A button appears in your workspace. You can click on this button to open TUTOR#6A. Move it to a convenient location (be careful not to click inside the outline of the button, or it will perform its function and open TUTOR#6A). Your workspace should look like Figure 6.12. Save the TUTOR#6B file.

Figure 6.12

- Click on the **menu button** to open TUTOR#6A.

File TUTOR#6B closes, and file TUTOR#6A opens. You can confirm this by pulling down the Window menu where all the open files are listed.

- Click on the **Window** menu to pull it down.

W̄ Only TUTOR#6A should be listed, as shown in Figure 6.13.

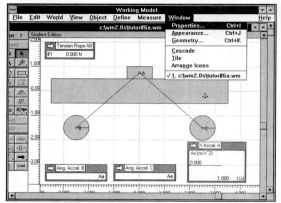

Figure 6.13

M The Window menu lists open file names only if more than one file is open at the same time. If no files appear, TUTOR#6A is the only file open.

You may need to resize your window to fit your workspace each time a new file opens.

- Create a similar **menu button** in TUTOR#6A that will **open TUTOR#6B** and position it conveniently.

Your workspace should now look like Figure 6.14.

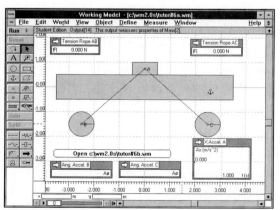

Figure 6.14

The Lock Controls option is similar to Lock Points but affects controls (meters, inputs, and menu buttons) rather than points. This option prevents mouse drags from changing the position of controls. Note that it also prevents you from opening any windows for controls; you must remember to unlock controls, for example, if you need to look at the Properties window for an output meter.

- Select **View** menu, **Lock Controls**.
- **Save** the **TUTOR#6A** file.

Switching between simulations

Now you can easily switch back and forth between the two simulations you have created.

- Click on the **menu button** to open TUTOR#6B.
- Select **View** menu, **Lock Controls**.

At this point the two files are identical; you will now change TUTOR#6B to fit the conditions laid out in the problem statement.

Splitting and Joining Objects

The Split tool "splits" joints, leaving a separate point on each mass object. You can edit these points individually, and then reassemble the joint with the Join tool. Elements that are split "remember" that they were once joined.

- Click on circle C.
- Click on the **Split** button on the tool bar.

The rope now appears as a dashed line, as shown in Figure 6.15.

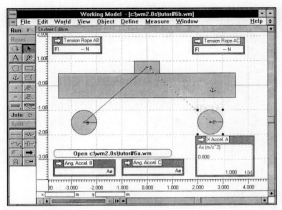

Figure 6.15

- Open the **Properties** window for circle C.

- If necessary, move the **Properties** window so that you can see circle C.

- Change the **y coordinate** of circle C to **-2**.

- Open the pull-down menu in the Properties window.

- Identify which point represents the end of the rope on circle C.

The correct point will have a * next to it. If you created your objects in the order that the tutorial instructed, this will be Point[9].

- Click on **Point[9]** in the pull-down menu.

- Change the **y coordinate** for this point to **0.5**.

- Close the **Properties** window.

This moves the point to the top of the circle; it is at the same height as the center point of body B, as shown in Figure 6.16.

You must now reconnect the joint (i.e., rope) between this circle and the slider.

- Click on the **Join** button on the tool bar.

The second experiment is now assembled as defined in the Problem Statement.

- **Save** TUTOR#6B.

Running the Simulation

- **Run** the simulation once, until both masses have returned to approximately where they started from.

Be sure to let the simulation run to at least Frame 191, because you will examine the results of this frame later in this tutorial.

Saving the Tape Recording

If you want to be able to observe your results without running the simulation each time you open the new file, use the Save tape recording option in the Save As... dialog box before switching. Using the Save tape recording option saves a much larger file that contains the simulation and the contents of the tape player; if you are short on disk space, you may want to forego this convenience and run the simulation each time you switch files. Using Save after you have saved the tape recording does not save the tape recording; you must reuse the Save As... option to save the tape recording again.

- Select **File, Save As...**

- Click **Save tape recording**.

- Select **TUTOR#6B** from the list that appears.

- Click **OK**.

- Click **Yes** to allow the previous copy of the file to be replaced.

- The default file name should be **TUTOR#6B;** click **Save** to accept it.

- Click **Replace** to allow the previous copy of the file to be replaced.

Figure 6.16

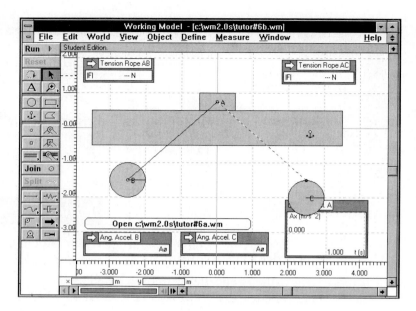

Now you will repeat the process with the other file.

- Open **TUTOR#6A** by clicking the **menu button**.
- **Run** the simulation again to at least Frame 191.
- Use **Save As…** with **Save tape recording** for TUTOR#6A.

Obtaining Results

- Use the **Tape Player** controls to move both simulations to **Frame 191**, the point where the circles appear to be at their maximum heights in TUTOR#6A.

Your simulations should look like Figures 6.17 and 6.18. It may be easier for you to jot down the meter readings for each of your simulations than to switch back and forth. In the symmetrical simulation, TUTOR#6A shown in Figure 6.17, the tensions in the two ropes are close to equal, the acceleration of the slider is approximately zero, and the circles have no angular acceleration. In the asymmetrical simulation, TUTOR#6B shown in Figure 6.18, the tensions are unequal, the slider is accelerating (and has moved from its original location), and circle C is rotating with an angular acceleration.

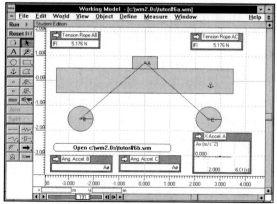

Figure 6.17

Figure 6.18

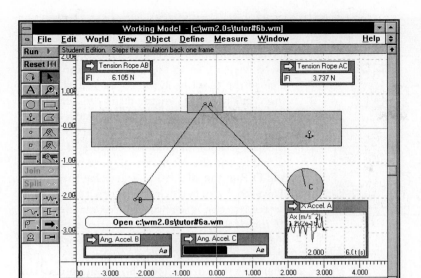

Changing the Scale of a Graph

Working Model dynamically scales graphical values as a simulation proceeds in order to obtain reasonable-looking graphs.

- Turn off **Lock controls** and open the **Properties window** for the **X Accel. A meter** in each of the experiments.

In the top half of the Properties window, the parameter y1 is defined as Mass[2].a.x, i.e., the acceleration of slider A in the x direction. The Min and Max scale factors for the parameter are shown in the bottom half of the window. In TUTOR#6A, -1.000 < y1 < 1.000, whereas in TUTOR#6B, -29.708 < y1 < 15.281. The Min and Max scaling values set by Working Model for y1 in Figures 6.19 and 6.20 emphasize what you have already observed—that circle C in the symmetrical simulation shows only some initial perturbation, whereas circle C in the asymmetrical experiment has a truly non-zero angular acceleration.

When comparing graphs, it is sometimes easier if they use the same scale. You can set the scale manually by removing the check mark (or X) from the Auto box to the left of the Min and Max values for y1. Enter the same maximum and minimum values for the variable in both meters and the graphs will appear with the same scale.

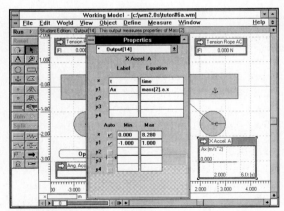

Figure 6.19

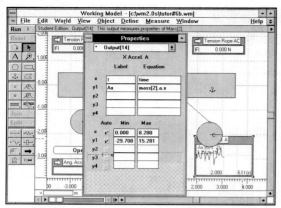

Figure 6.20

Link to Dynamics

The circles that are attached at their centers do not rotate; the circle attached at its perimeter swings about its point of connection and has an angular acceleration with respect to the fixed reference frame. The rotational inertia caused by the

angular acceleration of circle C causes the problem to be asymmetrical and gives slider A an acceleration. One result of this acceleration, by Newton's laws for rigid-body motion, is to change the tensions in the ropes. You can use the Working Model vector display feature to observe the dynamic effects more closely: Create vectors showing the force in both ropes, the acceleration of the mass centers of circles C and D, and the gravitational force on circles C and D. Run the simulation again. Use the tape player to move each circle, in turn, to a position where its rope is vertical (approximately frame 46 for circle B and frame 56 for circle C). Note that when rope AB is in the vertical position, the acceleration and forces on circle B are all in line; note the torque (couple) on circle C when rope AC is in the vertical position. What conclusions can you draw from these observations?

- **Save** (or use **Save As...** to keep the tape recording) your simulations.

- **Exit** Working Model.

Tutorial 6 • Exercises

6.1 The Basics Make a model that could simulate the possible instability of an overhead rolling crane system. Create an anchored rectangle that is 6 × 0.5 m to serve as the ground. Place a 1-m diameter circle on the ground, and place a 1-m square mass centered at a 45° angle from the circle below the ground, as shown in Figure E6.1. Connect the mass centers of the circle and the square with a rope. Create digital output meters for the circle's linear and angular velocity, a bar meter for the rope tension, and a menu button to run the simulation. Run the simulation and see if the information in the output meters makes sense. How are the linear and angular velocities of the circle related? Change the scale of the velocity output meters so that zero velocity is in the center of the graph. Split the rope's joint to the square, move the endpoint of the rope to the middle of the top of the square, and join the rope and square again. Run the simulation again and compare your results.

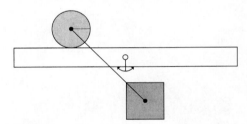

Figure E6.1

6.2 Applying the Basics Create a system like the one shown in Figure E6.2 to simulate the startup and motion of an overhead crane system. A 0.2 kg circle is on an anchored rectangle and connected with a rope to a 0.2-kg block (use the approximate location shown in the figure). The circle has an applied force of 2 N. Calculate the steady-state geometry of this system with free-body diagrams and Newton's laws. Create appropriate output meters to verify all of your calculations and run the simulation.

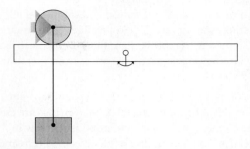

Figure E6.2

6.3 For Fun and Further Challenge Create a model of the system shown in Figure E6.3, which could be part of a positioning device in a conveyor system. Place 2 0.1-kg masses labeled A and B on an anchored surface. Connect the two masses with a rope and connect mass A to the ground with a force actuator. Set all friction to zero and enter the formula −cos(8*t), where t is time, for the actuator's force. Create output meters for the actuator's force, the rope's tension, and the acceleration of mass A. Run the simulation and see if the acceleration matches with the applied forces. Repeat for an actuator force of −2*cos(8*t). Change the settings so the masses will not collide and run the simulation again.

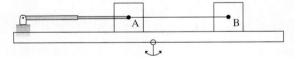

E6.3

6.4 Design and Analysis In your quest to design better vehicle suspensions, it becomes necessary to examine a model that can both translate vertically and rotate in the vertical plane much the same way that a car bounces up and down and rocks fore and aft. This type of model is called a pitch-heave (rocking-bouncing) model and can be implemented in Working Model as shown in

Figure E6.4. The vehicle body is the circle, A, and the simplified suspension is represented by the front and rear springs, S1 and S2 in the figure. The vertical, anchored rectangles limit the circle's motion to vertical translation and rotation if all friction is set to zero. The rope attached to square B is one way to represent an uneven mass distribution in the vehicle, since it is offset from the center of the circle.

Let mass A be 300 kg, mass B 30 kg, and give the springs a stiffness of 20,000 N/m. (Springs are created in the same way as actuators, as you saw in Exercise 5.4.) Set friction of all objects to zero and leave gravity on. Create output meters to graph the vertical and rotational motion of the vehicle body and the forces in the springs and run the simulation. Are the frequencies of heave and pitch the same? (That is, does the system bounce up and down as often as it rocks fore and aft?) How does varying spring stiffness change the motion? Move the springs closer to the center of the circle (use split and join) and compare the frequencies of translation and rotation.

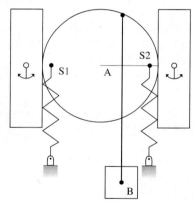

Figure E6.4

Exploring Your Homework These exercises involve pendulum-type systems where masses are constrained by ropes or cables.

6.5 The ball of mass m is released from rest in position 1. Determine the work done on the ball as it swings to position 2 (a) by its weight; (b) by the force exerted on it by the string. (c) What is the magnitude of its velocity at position 2? (d) What is the tension in the string in position 2?

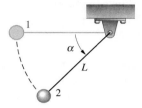

Figure E6.5

6.6 The 200-kg wrecker's ball hangs from a 6-m cable, as shown in Figure E6.6. If it is stationary at position 1, what is the magnitude of its velocity just before it hits the wall at position 2?

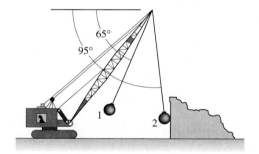

Figure E6.6

Exercises 6.5 and 6.6 are adapted from Bedford and Fowler's Engineering Mechanics: Dynamics *(Addison-Wesley, 1995), page 156.*

Impact and Energy

Objectives

In this tutorial you will learn to:
- Change numbers and units
- Create polygon objects
- Use the Help Ribbon
- Use the Point tool
- Join points to create joints
- Use a selection rectangle for selecting several objects
- Use nonstandard materials
- Change the coefficient of restitution
- Display contact forces
- Change the scale of meter graphs

Dynamics principles related to this tutorial

- Kinetic and potential energy
- Oblique impact
- Coefficient of restitution
- Impulse and momentum

Introduction

Work/energy and impulse/momentum approaches provide additional options for solving many engineering problems. In this tutorial, you will use Working Model to demonstrate some aspects of the principles of energy conservation and impulse and momentum. You will use the Polygon tool for the first time and you will join objects by forming joints. You will make extensive use of output meters and force displays.

Problem Statement

The system shown in Figure 7.1 is initially at rest. The ball of radius 3 in. is dropped from a height of 2 ft above the ground onto the wedge-shaped cart. The diameter of the cart's wheels is 1.5 in. Develop a Working Model simulation of this problem. Show graphs of the kinetic and potential energy of the ball and the kinetic energy of the wedge as functions of time. Display all contact forces. Assume that the cart does not lose contact with the ground, friction effects are negligible, all objects are made of steel, and all collisions involving the ball are perfectly elastic.

Figure 7.1

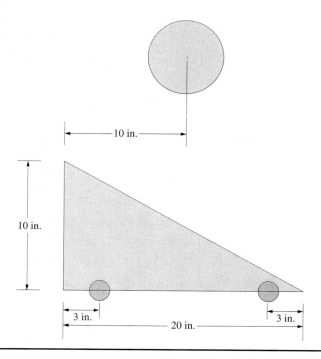

Setting Up the Workspace

- **Open** a new, untitled Working Model window.

Your workspace should open with x,y axes, grid lines, rulers, and coordinates displayed in the workspace. If they are not visible, turn them on.

You should be using the Working Model default for accuracy, Accurate mode.

Changing the Numbers and Units

Your new simulation uses the default Working Model unit system, which is SI (degrees), the SI/metric unit system. You can change this to fit your needs by using the Numbers & Units dialog box.

- Select **View** menu, **Numbers & Units...**

The Numbers & Units dialog box appears on your screen, as shown in Figure 7.2.

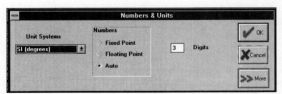

Figure 7.2

You can change the display of units within Working Model to a variety of unit systems: Astronomical, Atomic, CGS, Custom, English (Earth pounds), English (slugs), SI (degrees), and SI (radians). You can choose how num-

bers are displayed by selecting Fixed Point, Floating Point, or Auto in the Numbers field of the Numbers & Units dialog box. Fixed-point format displays all numbers with a fixed number of digits to the right of the decimal point. Floating-point format displays all numbers in an exponential format of the form 1.23e4. The default option, Auto, lets Working Model decide whether to display numbers in fixed- or floating-point format. Auto is usually the best choice. You can also select the number of digits that Working Model displays in your simulation.

This problem is given in English units. You will now change the Working Model units to the English system. You will change the number of digits displayed during the simulation from 3 to 4. This will make the conversion factors that Working Model uses internally more accurate. This is important to remember—if you reduce the number of digits and use units other than SI, you can seriously affect the accuracy of your simulation. You will change the Numbers display to fixed to make it easier to set up your components. (Working Model's Auto setting uses floating point notation when more than 3 digits are displayed.)

- Click on the **Unit Systems** pull-down menu.

- Choose **English** (**slugs**).

- Click **Fixed Point** in the **Numbers** area.

- Change the 3 in the **Digits** field to **4**.

W Click the **More** button.

M Click the **More Choices** button.

Your dialog box should resemble Figure 7.3.

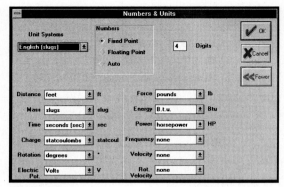

Figure 7.3

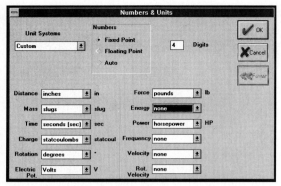

Figure 7.4

All the units change to the English system; however, some of them may still not be exactly what you want. For example, the measurements in the Problem Statement are in inches; setting up your simulation will be easier if you change the distance units from feet to inches. In addition, Working Model defaults to Btu for its English energy units; for this problem it is preferable that energy be expressed in lb-in. If you select the option none under Energy, Working Model will use the units that result from force × distance for the output meters (in this case, lb-in.).

- Click on the **Distance** pull-down menu.

- Select **inches.**

- Click on the **Energy** pull-down menu.

- Select **none.**

When any of the specific measurement variables for a certain type of unit system are changed, as you have changed the distance and energy options, the Unit Systems name in the dialog box changes to Custom, as shown in Figure 7.4.

- Click **OK.**

Note that coordinates and rulers now measure everything in the workspace in inches, as shown in Figure 7.5.

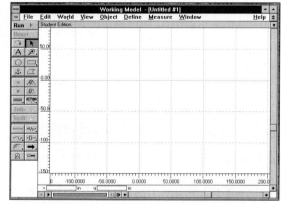

Figure 7.5

If you wanted to use these numbers and units as the default for all new simulations, you would select Save Current Settings from the Preferences dialog box. Don't do this now; the rest of the tutorials assume you will use the workspace settings saved in Tutorial 1.

Creating the Components

The wedge is 10 in. high and has a 20-in. base. You will now change the view size of the workspace to more suitable dimensions so that your model will appear large enough in the workspace. If you need help with View Size, refer to Tutorial 4.

- Select **View** menu, **View Size...**
- Change the dialog box to read **Objects on screen are 0.1 times actual size.**
- Click **OK.**

Creating Polygon Objects

The *Polygon* tool is used to create irregularly shaped mass objects. To use the Polygon tool, you define each point with a single click and double-click to signal the last point that defines the shape. The polygon automatically closes, connecting the point you double-click with the first point. Using the Polygon tool and the screen coordinates, you will now create, place, and size the wedge-shaped body.

While you are creating an object, the coordinates information shows distance relative to the last point you selected (how far your cursor has moved). The distance is shown in terms of x and y values, Δx and Δy, as well as total distance from the last point and displacement angle. When you have completed the polygon, the coordinates will return to the absolute x and y coordinates display of the location of the cursor.

Tip The Coordinates display also changes from absolute coordinates to relative coordinates when you drag an object.

- Click once on the **Polygon** icon (to the right of the anchor).
- Click once at the **origin** in the workspace.
- Click at the relative coordinates, $\Delta x = 20$, $\Delta y = 0$.
- Double-click at the relative coordinates, $\Delta x = -20$, $\Delta y = 10$.

Your workspace should look like Figure 7.6.

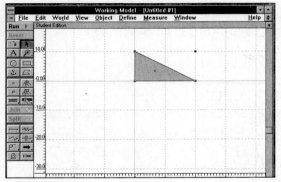

Figure 7.6

Now you will create the ball and the wheels of the cart.

- Create three **circles** in the workspace.

One of these circles will be the ball and the others will be the wheels on the wedge. Using the Geometry window,

- Make the radius of the ball **3.0 in.**
- Make the radii of the wheels **0.75 in.**

Your screen should look like Figure 7.7.

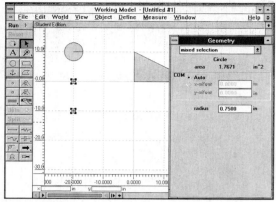

Figure 7.7

- Using the Properties window, place the center of the ball at the coordinates (**10,24**).

The ball moves out of view. Your workspace should look like Figure 7.8.

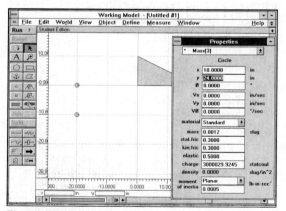

Figure 7.8

- Close the **Properties** and **Geometry** windows.

Using the Help Ribbon

The Help Ribbon displays information about the tool or object that is currently under the pointer. For example, if you move the cursor over the wedge in your workspace, the Help Ribbon identifies it as Mass[2] Polygon (your mass number should be the same if you created your components as instructed). This is a good way to determine the mass number of objects. If the cursor is over a tool on the tool bar, the Help Ribbon identifies it.

Using the Point Element

You have seen how you can use the *point element* for measuring properties at a specific location on a mass object, as you did with the rolling contact problem in Tutorial 5. You can also combine a point with a slot or another point to form a joint. (You will learn more about slot joints in Tutorial 11.)

Don't confuse the round Point tool with the square Point tool just below it. The Help Ribbon in Figure 7.9 shows that the cursor is pointing to the round Point tool. You use the square Point tool to form rigid (nonturning) joints, and the round Point tool to form pin (turning) joints.

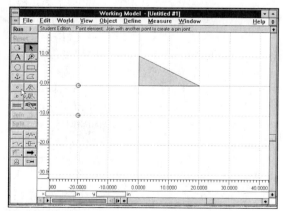

Figure 7.9

You will create four round points and then join them into two pairs of pin joints. The exact locations of the new points are not important yet. The most important thing is that you place the points *on* the cart and *on* the wheels, and away from their edges so that they do not inadvertently attach to the background. You will locate these points accurately afterwards, using the Properties window. As long as you attach the point to an object, you can move it anywhere on the object using the Properties window; if you attach a point to the background, you must first drag it to attach it to the object before you can move it to its proper location, using the Properties window.

- Double-click on the **Point** icon.
- Click at the center of each wheel.
- Click in two different locations on the wedge.
- Click on the **Arrow** icon to turn off Point.

Your workspace should look like Figure 7.10.

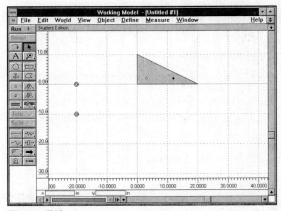

Figure 7.I0

- Click on the point on one of the wheels.
- Open the **Properties** window for the point.

Remember that the coordinates indicated in the top half of the Properties window are relative to the mass center of the object to which the point is attached. The points on the wheels should be located at the wheels' mass centers. Using the Properties window,

- Check to see if the coordinates of this point are both **0**. If not, change them to **0**.
- Make sure the point on the second wheel is also located at (**0,0**) on the wheel.
- Click on the leftmost point located on the wedge.

You may need to move the upper portion of the Properties window or scroll your workspace to see the wedge and its connected points more completely.

The x and y values in the upper portion of the Properties window show the coordinates of the point relative to the mass center of the wedge. You could locate the points that will be used to attach the wheels to the wedge by calculating their distance from the wedge's mass center. Since the problem states that these two points must be located at 3 in. from either end of the wedge, and the global coordinates of the wedge are known, it is easier to place the wheels using the global coordinates at the bottom of the dialog box. (Since the wedge's mass center is located at (6.6667, 3.3333) the wheels' relative coordinates must be (-3.6667,-3.3333) and (10.3333,-3.3333), respectively. Check this as you locate the points.)

- Change the leftmost point's **global x** and **y** values to **3** and **0**, respectively.
- Click on the rightmost point attached to the wedge.
- Change its **global x** and **y** values to **17** and **0**.
- Close the **Properties** window.

Your workspace should look like Figure 7.11. Lines may appear from the wedge's center of mass to each of the points you just relocated. When points are attached to a mass, then moved off the mass, these lines appear to indicate that the points are still attached but no longer positioned on the mass. Because your points are so close to the edge of the wedge, these lines may appear.

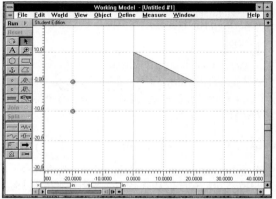

Figure 7.11

Joining Points to Create Joints

The Join button can be used to form a joint between two elements. To use this tool, select both of the elements and then click the Join button. The original positions of elements need not overlap. Joining two round point elements forms a pin joint; joining two square elements forms a rigid joint.

In this case, you want to form pin joints that cause the wheels to move over to the wedge, and not the wedge to move to the wheels. To force this to happen, you will temporarily anchor the wedge.

■ Click on the **Anchor** icon.

■ Click on the wedge.

The wedge is now anchored and will not move. Next you will join the wheels to the cart.

■ Click on one of the points on the wedge.

■ Hold the **Shift** key down and click on the center point on one of the wheels.

Your workspace should look like Figure 7.12. If you accidentally selected the wheel, deselect it and try again. Only the two points should be highlighted.

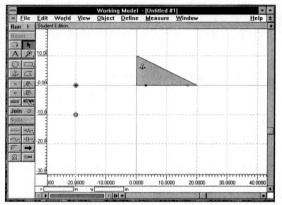

Figure 7.12

■ Click the **Join** button.

■ Repeat to join the other wheel to the cart.

You can remove an anchor from an object by placing another anchor on it (two anchors cancel each other out).

■ Click on the **Anchor** icon.

■ Click on the wedge.

> **Tip** You can also remove an anchor from an object by selecting the anchor and choosing **Edit** menu, **Clear**. **W** The Del key is the keyboard command for clearing an item in Windows.

Your workspace should look like Figure 7.13.

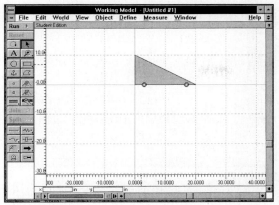

Figure 7.13

Using a selection rectangle

You can select objects near each other by enclosing them in a selection rectangle. You will now select the cart and wheels assembly, using a selection rectangle. As you are selecting your area, Working Model displays a dashed rectangle to indicate the selected area. When you release the mouse button, all objects enclosed by the rectangle are selected. All other objects are deselected.

> **Tip** Holding down the Shift key while dragging the selection rectangle toggles the selection state of all enclosed objects: objects that were previously not selected become selected, and objects that were previously selected become deselected.

- Click on the background above and to the left of the cart.

- Drag to the lower-right corner below the cart.

Your screen should look like Figure 7.14.

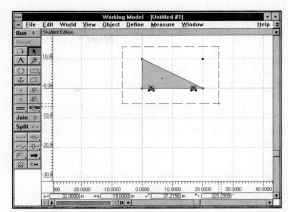

Figure 7.14

- Release the mouse button.

All the parts of the cart and the wheels assembly should be selected, as shown in Figure 7.14.

- Move the entire wedge assembly by clicking and dragging on any item in the selected area, so that the wheels rest on the horizontal x axis.

The left side of the wedge must continue to touch the y axis (if you have trouble aligning the assembly, temporarily click off Grid Snap on the View menu).

- Deselect the objects, then use the scroll bars to make your workspace look like Figure 7.15.

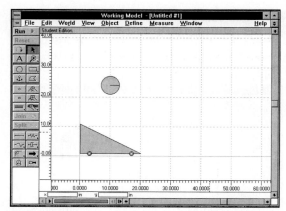

Figure 7.15

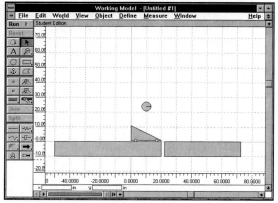

Figure 7.16

You will now create the ground surfaces on which the cart will roll and the ball will bounce. The ground must be represented by two solid objects: one that the cart will roll on, and one that the ball will bounce on. This is necessary because the cart does not lose contact with the ground (i.e., the contact will be perfectly plastic), but the ball will bounce on the ground (it is defined as perfectly elastic).

- Turn **Grid Snap** back on, if necessary.

- **Zoom out** once on the workspace.

- Create one ground **rectangle** with opposite corners at coordinates (**-50,-10**) and (**20,0**).

- Create the second ground **rectangle** with opposite corners at coordinates (**22.5,-10**) and (**72.5,0**).

Your workspace should look like Figure 7.16.

- **Anchor** both ground rectangles.

- **Zoom in** and use the scroll bars to enlarge your workspace and move the screen so that the cart is at the bottom left corner.

Your workspace should look like Figure 7.17.

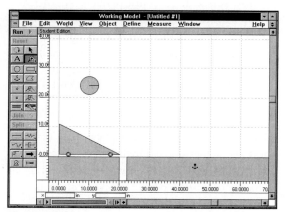

Figure 7.17

Setting Material Properties

You can easily set many of a mass object's properties to reflect a specific type of material. These settings are approximate, and give typical values of properties for materials such as rubber, rock, plastic, ice, clay, wood, and steel. Picking a material sets a mass object's mass, coefficients of friction, coefficient of restitution (elasticity), charge, density, and moment of inertia. You can change any of these values to customize a particular application. For this problem, all the items are made of steel, but the ball, wedge, and right-hand ground are elastic, whereas the wheels and left-hand ground are plastic.

- Select **Edit** menu, **Select All.**

- Double-click on any item to open the **Properties** window for this mixed selection.

This Properties window is for all objects, since everything in your workspace has been selected.

- Click on the pull-down menu for **material.**

Your materials selections should appear as shown in Figure 7.18.

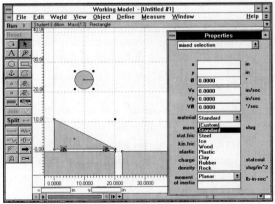

Figure 7.18

- Select **Steel.**

Since the friction effects are defined in the problem statement as negligible, you will eliminate friction for all objects.

- Change **stat. fric** to **0.**

- Change **kin. fric** to **0.**

- Click on a blank area of the workspace to deselect everything.

Changing the Coefficient of Restitution

The coefficient of restitution, e, is a measure of how much energy is lost during an impact. $0 < e < 1$ represents its range of values, where $e = 1$ is for a "perfectly elastic" impact in which no energy is lost, and $e = 0$ is for a "perfectly plastic" impact in which the two objects remain stuck together. The coefficient of restitution depends on the materials and geometries of the colliding bodies. In Working Model, e is represented by the number in the elastic box in the Properties window.

- Use the **Shift** key to select the ball, the wedge, and the right-hand ground rectangle.

- In the Properties window, change the value in the **elastic** text box to 1.

- Click on a blank area of the workspace to deselect everything.

- Use the **Shift** key to select the wheels and the left-hand ground rectangle.

You can use the Help Ribbon to tell you when the cursor is over the wheels and not the pin joints.

- In the Properties window, change the value in the **elastic** text box to **0.**

As with the unit systems in the Numbers & Units dialog box, when you change any of the material properties for a certain type of object, as you have changed the elasticity value of steel, the material name in the Properties window changes to (Custom), as shown in Figure 7.19.

The Properties described by charge, density, and moment of inertia will be discussed in Tutorial 15.

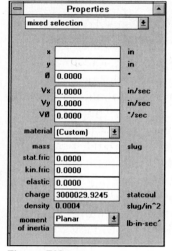

Figure 7.19

- Close the **Properties** window.
- Use the Appearance window to label the ball with the letter A, the wedge with the letter B, and the right-hand ground with the letter C.
- **Save** your simulation now as **Tutor#7.**

Your workspace should look like Figure 7.20.

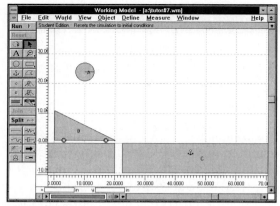

Figure 7.20

Running the Simulation

- **Run** the simulation until the ball has bounced twice, once on the wedge and once on the right-hand ground.

- Click **Reset.**

If you see the message "Objects are overlapping beyond the specified tolerance," your wheels are overlapping the ground rectangle. Click Stop and move the wedge and wheels assembly up a bit. Zooming in can help with this positioning.

If the wedge and wheels assembly bounces on the ground, then the wheels were not touching the ground. Select the whole assembly and move it slightly closer to the ground.

Tip One way to position the wheels and wedge assembly accurately is to let the simulation run until the assembly settles on the ground. Use the Tape Player to go to the first frame of the simulation where it is in position, then select Start Here from the

World menu to reset the initial conditions. Using this method in this simulation, you would need to reposition the ball after getting the wheels and wedge assembly in place.

Notice that the ball does not appear to make contact with ground C. You need to make the animation step smaller so you can see the contact, as you did in Tutorial 4.

- Select **World** menu, **Accuracy...**
- In the **Animation Step** box, click the button below Automatic.
- Change step size to **0.01** s.
- Click **OK**.
- **Run** the simulation again.

See if Working Model shows the frame at which the ball bounces on ground C. Make sure that any vertical momentum of the wedge assembly has been eliminated. Turn Tracking off if it is on.

Obtaining Results

- Create three output **meters** in **graph** display:
  1. The **translational** and **rotational kinetic energy** of ball **A**
  2. The **gravity potential** (potential energy) of ball **A**
  3. The **translational** and **rotational kinetic energy** of cart **B**

Using the **Appearance** window,

- Put **grids** and **axes** on all the graphs.
- Label the meters appropriately and arrange them conveniently in your workspace.

Tip To create a translational and rotational kinetic energy meter, create one meter for all kinetic energy for the object, and then display only the translational and rotational outputs.

Your workspace should look like Figure 7.21.

Figure 7.21

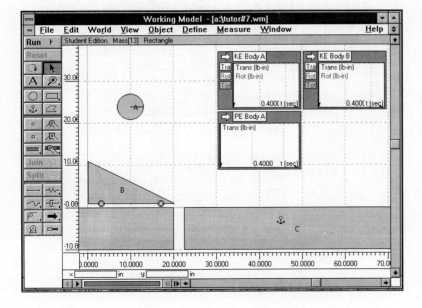

Note that the units for energy appear as lb-in. Remember that you selected none as the unit for energy.

- Create a temporary **meter** that shows the **time.**

  Tip: The **Time** meter may open in the workspace behind the left-hand ground rectangle. If this happens, scroll the workspace so that you can reposition the **Time** meter, then scroll the objects back into place.

- **Run** the simulation to see how long it takes until the ball has finished its second impact.

This impact occurs at approximately 0.65 sec, as shown in Figure 7.22.

- **Reset** the simulation.

- Clear the **Time** meter from your screen.

Changing the Meter Graph Scale

Working Model automatically sets and changes the scale of the graphs in your output meters as a simulation runs. The scale for each graph may be different, and you may control it by overriding the default settings. By changing the scale, you can make it easier to compare the graphs of different variables, and you can restrict the graphic display to the time period you are interested in. In this example, you will focus the graphs on the time period until just after the second impact by manually inputting the range for the meters' time scale as .70 seconds.

- Select all the meters.

- Open the **Properties** window.

- Click the **Auto** button off next to the first variable, **x**, in the lower half of the window.

Figure 7.22

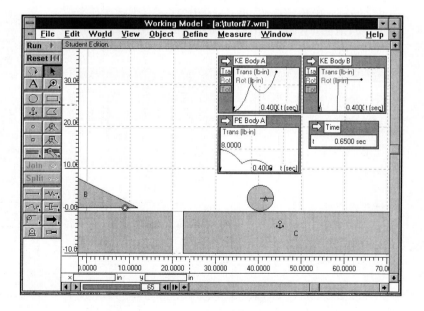

Refer to Figure 7.23.

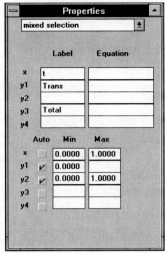

Figure 7.23

- Change the **Max** value on the first variable, **x**, to **.7**

Now all the meters have the same time scale.

- Deselect all the meters.

- Click on one meter at a time to look at the Min and Max values for each graph.

The values displayed will be similar to those under "Original" in Table 7.1. They may not be exactly the same in your simulation, since you may not have stopped the simulation at exactly the same point in time. (Remember that Working Model recalculates the range and scales the graph to accommodate the simulation.) Change these values to match the ones under "New" in Table 7.1, so that results can be more easily interpreted. Figure 7.24 shows the Properties windows for the kinetic energy of body A after the changes to the new y values.

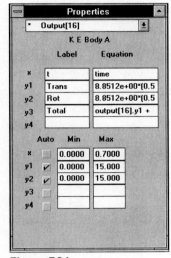

Figure 7.24

- Close the **Properties** window.

- **Run** the simulation again to see the difference in the meters' display.

	Original Min	Original Max	New Min	New Max
K E Body A - translational	0.00	11.95	0.00	15.00
K E Body A - rotational	0.00	1.00	0.00	15.00
P E Body A	0.00	18.18	0.00	15.00
K E Body B - translational	0.00	1.00	0.00	1.00
K E Body B - rotational	0.00	1.00	0.00	1.00

Table 7.1

Displaying contact forces

The problem statement stipulates that your simulation should show all the contact forces as they occur. To display the vectors,

- Select all the mass objects by using the **Select All** option then holding the **Shift** key down while deselecting the three meters.

- Select **Define** menu, **Vectors, Contact Force.**

- **Run** the simulation again.

Scaling vectors

Some of the vectors appear to be very long, as shown in Figure 7.25; you will need to scale them as you did the velocity vectors in Tutorial 4.

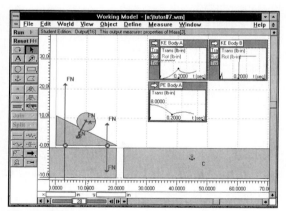

Figure 7.25

- **Reset** the simulation.

- Select **Define** menu, **Vectors Lengths...**

- Change the value under the **Force** slider to **.001.**

- Click **OK.**

- **Run** the simulation again.

The vectors now appear shorter.

- Use the **Tape Player** to move the simulation to the point where the ball strikes the wedge.

Your workspace should look like Figure 7.26.

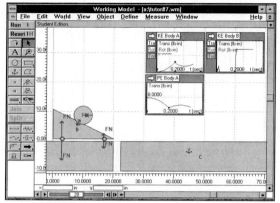

Figure 7.26

Note the exchange of kinetic and potential energy for the ball as shown in the graphs. After the first impact, the kinetic energy for the wedge remains at a constant low value. (Remember, you selected a larger scale for this graph when you entered a maximum of 1.) Notice, too, that the contact forces are always normal to the surfaces.

Link to Dynamics

Is linear momentum conserved for the ball-wedge-wheel system? During the first impact, there are external forces (or impulses) acting between the ground and the wedge-wheel system in the y-direction; this means that linear momentum will only be conserved in the x-direction. Linear momentum for the ball-wedge-wheel system is, therefore, not conserved along the line of impact or perpendicular to it, because the vertical forces have components in these directions. However, for the ball *alone*, linear momentum is conserved in the direction perpendicular to the impact, since no external force acts in this direction.

Is energy conserved throughout the motion? Since the effects of friction are assumed to be negligible, and the impacts are elastic, there should be no energy loss. You can use work-energy equations to verify this and to check the accuracy of your solution. You can open an additional output meter to easily obtain the height of the ball at various times, and the masses of the moving objects can be found in the Properties window.

- **Save** the simulation and **exit** Working Model.

7.1 The Basics Set up the workspace with grid, rulers, and axes. Change the units system to English (Earth pounds) and set the view size to 0.1. Put together an anchored ground, a double-wedged cart with wheels, and a circular mass as shown in Figure E7.1. Use pin joints for the wheels as you did in Tutorial 7. Set the material for the ground to rock; the cart, wood; the wheels, steel; and the ball, rubber. Show contact forces with properly scaled vectors. Run the simulation. Create meters to examine conservation of energy and momentum. Adjust the coefficient of restitution so all collisions are perfectly elastic and run the simulation again. Can you explain the difference between the two simulations?

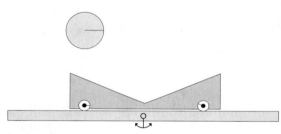

Figure E7.1

7.2 Applying the Basics The manufacturing industry makes use of many conveyors. Create a conveyor by using pin joints to attach 8 identical rollers to an anchored base. Add a 1-kg square mass that is sized so that it is always supported by at least 2 rollers, as shown in Figure E7.2. (Don't make it so narrow that it could wobble on one roller alone.) Use the Force tool to apply a 10-N force to the mass and define the force so that it is only active when the mass is on the left half of the conveyor. (Enter a formula in the Active when field of the force's Properties window. Mass[1].p.x is the variable name for the x-position of the mass center of Mass[1].) Set up meters for the velocity of the square mass and the angular (rotational) velocity of

the fourth roller from the left. See if you can explain why the conveyor-mass system behaves the way it does. What happens if you change the frictional characteristics of the rollers and mass?

Figure E7.2

7.3 For Fun and Further Challenge Create a small billiard game with four circles: three circles placed together and one at a distance, as shown in Figure E7.3. Build the table bumpers using the Polygon tool. Turn gravity off and make all collisions perfectly elastic. Give the single ball an arbitrary initial velocity and monitor the momentum and kinetic energy of the system before and after the collisions.

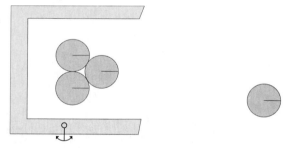

Figure E7.3

7.4 Design and Analysis In the power train department at Awesome Auto, you need to study accelerating ground vehicles. A scaled model will do for the first cut. Build a car with wheels running on the ground as shown in Figure E7.4. The car has a mass of 10 slugs and the wheels are 0.1 slug each. Add a force acting at the car's mass center that represents the traction force of the car. Position the base of the force using the Properties window. Define the force as a function of the car's velocity with the formula −100−mass[1].v.x (where the car is Mass[1]). This will define the force to be larger at low velocities and smaller at high velocities, which is realistic given air resistance, friction, and engine properties. Run the simulation. Monitor the force and velocity of the car, the angular velocity of the wheels, and the contact forces. Relocate the applied force to first the top and then the bottom of the car and re-run the simulation. What happens to the wheel contact forces?

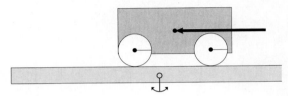

Figure E7.4

Exploring Your Homework   The following problems examine the relationship between external forces and the changes of momentum in systems.

7.5   A person of mass m_P stands at the center of a stationary barge of mass m_B, as shown in Figure E7.5. Neglect horizontal forces exerted on the barge by the water. If the person starts running to the right with velocity v_P relative to the water, what is the resulting velocity of the barge relative to the water?

Figure E7.5

7.6   During the first 5 s of the 32.2-kip airplane's takeoff roll, the pilot increases the engine's thrust at a constant rate from 5 kip to its full thrust of 25 kip. What impulse does the thrust exert on the air-

plane during the 5 s? If you neglect other forces, what total time is required for the airplane to reach its takeoff speed of 150 ft/s?

Figure E7.6

7.7   A bullet (mass m) hits a stationary block of wood (mass m_B) and becomes embedded in it. The coefficient of kinetic friction between the block and the floor is μ_k. As a result of the impact, the block slides a distance D before stopping. What was the velocity v of the bullet? (Strategy: First solve the impact problem to determine the velocity of the block and the embedded bullet after the impact in terms of v, then relate the initial velocity of the block and the embedded bullet to the distance D that the block slides.)

Figure E7.7

Exercises 7.5, 7.6, and 7.7 are adapted from Bedford and Fowler's Engineering Mechanics: Dynamics *(Addison-Wesley, 1995), pages 191, 196, 204.*

Exploding Shell

Objectives

In this tutorial you will learn to:
- Create spring elements
- Describe forces with polar coordinates
- Use the Active when field
- Display the system center of mass
- Create new reference frames
- Customize the parameters in an output meter

Dynamics principles related to this tutorial

- Motion of the mass center of a system of particles
- Linear momentum of a system of particles
- Kinetic energy for a system of particles

Introduction

In this tutorial you will create a simulation for a type of problem known as an *exploding shell* problem. An exploding shell is a system that starts as a rigid body, but after a certain time breaks into several pieces. The mass center of the pieces behaves as if the system were still in one piece. For example, the famous Shumacher-Levy comet, which was broken into pieces by the planet Jupiter's atmosphere, continued to orbit Jupiter as a group of particles for years until it crashed into the planet in July, 1994.

Exploding shell problems have noncontinuous conditions; in this tutorial, you will learn how to deactivate forces and springs at specified times. You will create a new reference system from which to observe the motion, and you will trace the path of the system mass center and learn how to customize output results.

Problem Statement

A fireworks manufacturer uses many different techniques to produce spectacular effects. In one, three packages of explosives are attached to each other by springs and propelled into the sky. In order to produce different effects, electronic devices are used to deactivate the springs at specific times before the fireworks are ignited. Figure 8.1 represents a system of explosives and springs consisting of 3 circular masses of radius 0.5 m, connected by 3 springs ($K = 100$ N/m) whose free lengths are all 0.1 m. An initial force, $F = 180$ N, is applied to mass A, as shown. At $t = 0.06$ s, the force is removed.

Create a Working Model simulation to display the path of the mass center of the system for the following two cases: (1) the springs never break and the system moves as a single rigid body; and (2) at $t = 0.1$ s, springs AB and BC break, and at $t = 0.6$ s, spring AC breaks. For Case 2, create a reference frame at the mass center of the system, and display the motion of the system with respect to that reference frame. Also, create an output meter showing the total linear momentum of the system. Is the momentum of this system conserved?

Figure 8.1

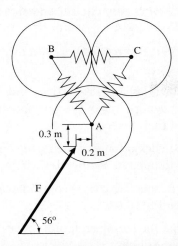

Starting Working Model

- **Open** a new, untitled Working Model window.

Your workspace should have rulers, grid lines, and axes as before. You should be using the Working Model defaults for Numbers and Units (SI (degrees)) and Accuracy (Accurate mode).

Creating the Components

Use Figure 8.1 as your guide as you set up the components of your model.

- Create three **circles** with radii of **0.5 m** in the workspace.

- Use the **Appearance** window to label the circles **A, B,** and **C.**

- Use the **Properties** window to place circle A at (**0,0**), circle B at (**-0.5,0.866**), and circle C at (**0.5,0.866**).

- Close the **Appearance** and **Properties** windows.

Your workspace should look like Figure 8.2.

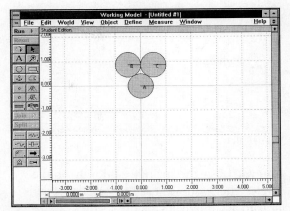

Figure 8.2

Creating and assigning properties to spring elements

Working Model springs are constraints that resist stretching or compression. You can attach *springs* between one mass object and the background, or between two mass objects (the endpoints of the spring are the attachment points). If you click and hold on the Spring tool on the menu bar, you will see two other types of spring elements, rotational springs and spring dampers, as shown in Figure 8.3. *Rotational springs* produce a twisting force as they are wound up. You can place them on top of a single mass object, in which case they will connect that object to the background, or you can place them on two overlapping objects to connect them. They have a built-in pin joint. *Spring dampers* are a combination spring and damper. They can be attached between one mass object and the background, or between two mass objects, in the same way regular springs are.

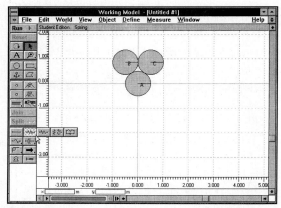

Figure 8.3

You will place the springs randomly on the circles; then you will use the Properties window to accurately locate the springs.

- Double-click on the **Spring** icon on the tool bar (the right-hand icon below the Split button).
- Click and hold the mouse button at any location on circle B.
- Drag to any location on circle A.
- Release the mouse button.

Your workspace should look like Figure 8.4.

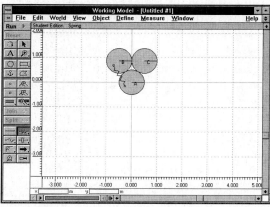

Figure 8.4

- Place two more **springs** on the circles to connect circle A to circle C and circle B to circle C.
- Click on the **Arrow** icon to deactivate the spring.

Now you will use the Properties window to line up the endpoints of the springs with the centers of their respective circles.

- Double-click on one of the spring endpoints to open its **Properties** window.
- Change the relative (x,y) coordinates for this point to (**0,0**).

Notice that the spring endpoint moves to the center of the circle.

- Repeat this for all the endpoints of the springs.

Tip A shortcut for changing multiple points or objects to the same setting is to select all the points or objects at once (by holding down the Shift key and clicking on them) and then change the appropriate setting. In this case, select all the spring endpoints and change the relative (x,y) coordinates in the Properties window for the entire mixed selection to (0,0).

Your workspace should look like Figure 8.5.

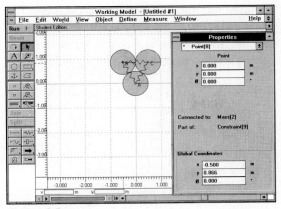

Figure 8.5

Take a look at the properties for each individual spring, as shown in Figure 8.6. Notice that the length is 0.958 m; yours may be different, depending on the length of the random springs you created. You can see "(current) 1.000 m" directly under the length box. The length box indicates the free (unstretched/ uncompressed) length of the spring. The (current) length indicates the current stretched or compressed spring length. When you create the springs, the two lengths are equal by default. The current length is determined by the location of the endpoints; you can change the free length in the Properties window.

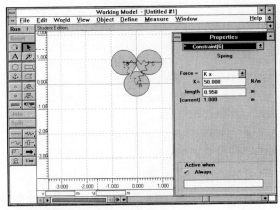

Figure 8.6

- You will now change the spring properties, using the mixed selection shortcut.
- Select all the springs.
- Change the **spring constants** to **K = 100** in the Properties window.
- Change the **spring lengths** to **0.1.**

You can change the current length without changing the free length by clicking and dragging one endpoint of the spring while holding down the **W** Ctrl key or **M** Command key. Try doing this with the lower point of spring AB.

W Hold the **Ctrl** key down and select point **A** on spring AB.

M Hold the **Command** key down and select point **A** on spring AB.

- Drag to anywhere on the screen.
- Select spring **AB.**

Figure 8.7 shows what your screen should look like. Observe that the changes appear in (current) length; length remains the same. Use Edit menu, Undo Drag to return the point to where it was originally.

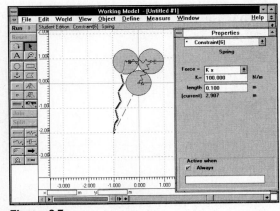

Figure 8.7

- Select **Edit** menu, **Undo Drag.**

Your workspace should look like Figure 8.8.

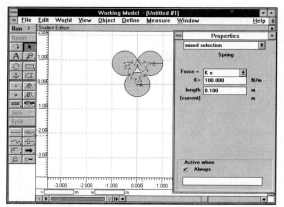

Figure 8.8

- Close the **Properties** window.

Describing forces in polar coordinates

You are now going to create a force on circle A, using Figure 8.9 as a guide. If you need help creating forces, refer to Tutorial 3.

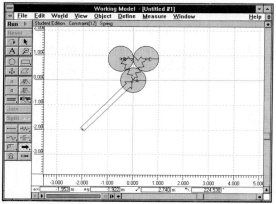

Figure 8.9

- Click on the **Force** icon on the tool bar.
- Click and drag from circle A down and to the left to create a force.

Note: You don't need to be particularly careful about the placement and size of the force; you will set its exact location and magnitude from the Properties window.

- Double-click on the force to open its **Properties** window.

The Properties window shows the force in rectangular (Cartesian) components by default.

- Click the radio button for **Polar** in the force's Properties window.

The Properties window for the force on your screen changes to resemble Figure 8.10 (your exact values may be different, depending on how you created your force).

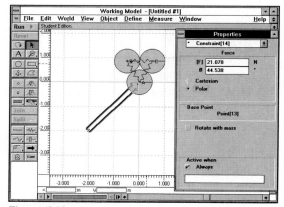

Figure 8.10

Now the components of the force are displayed in polar coordinates; the top text box under Force shows the magnitude of the applied force, and the text box below it shows the angle in degrees.

- Change the value of the force magnitude, |**F**| (in the upper text box), to **180** N.
- Change the **angle** in the lower text box to **56°**.

You would like the force vector to appear shorter in the workspace. You will therefore adjust the scale for the length of the vector.

- Select **Define** menu, **Vector Lengths...**
- Change the value under the **Force** slider to **0.01.**

Remember from Tutorial 4 that the value in the Vector Lengths dialog box × the magnitude of the force will yield the resulting length of the vector. In this case, 0.01 × 180N results in a vector that is 1.8 units long. Your dialog box should look like Figure 8.11.

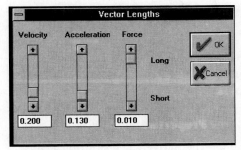

Figure 8.11

- Click **OK.**

The new scale of the force vector makes it easier to view in the workspace, as shown in Figure 8.12.

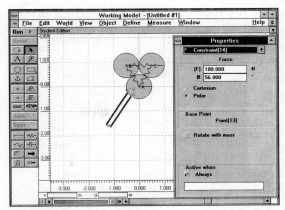

Figure 8.12

Now you will select the point of application, or base point, of the force. The Properties window tells you which point the force is applied to in the area under Base Point. In Figure 8.12, it is Point[13]; yours may be different. Choose the point that is serving as the Base Point by using the pull-down menu at the top of the Properties window; you could select it with the mouse in the workspace, but this is more difficult. As the problem states, the force is applied on circle A at a specific point below and to the left of the circle's center.

- Select the **base point** of the force.
- Locate the base point at (**-0.2,-0.3**) with respect to the mass center of circle A.

Your workspace should look like Figure 8.13.

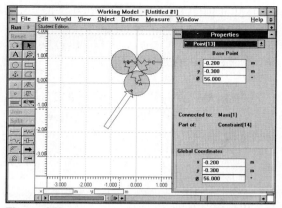

Figure 8.13

When you have finished placing all your points and objects, you will use Lock Points to make sure that you do not move any of your carefully selected locations by mistake.

- Select **View** menu, **Lock Points**.

Using the Active when Field

In Working Model, each constraint can be turned on and off during the course of a simulation. The bottom section of the Properties window for all constraints contains a field called Active when. This field is set to Always by default, as shown in Figure 8.14, but it can be changed. According to the Problem Statement, the force in this simulation is to be active until $t = 0.06$ s.

- Select the force.

W Click the **Always** button **off** in the Active when field of the force's Properties window.

M Click the button below **Always on** in the Active when field of the force's Properties window.

- Enter **t<.06** in the Active when text box.

Note: In the future, you will simply be directed to change the condition in the Active when field; individual Windows and Macintosh instructions will not be provided.

Your screen should look like Figure 8.14.

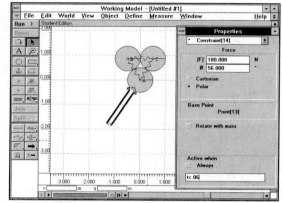

Figure 8.14

- Close the **Properties** window.

You will use the Active when option to cause the springs to break during the simulation later in this tutorial.

Displaying the system center of mass

- Select **View** menu, **System Center of Mass**.

There is no visible change in the workspace yet. The system center of mass indicator will not appear until the simulation is run.

- Click the **Run** button.

- When the **x** appears in between the three circles, click **Reset**.

The **x** indicates the system mass center, as shown in Figure 8.15. The Problem Statement requires that you trace the path of this point during the simulation.

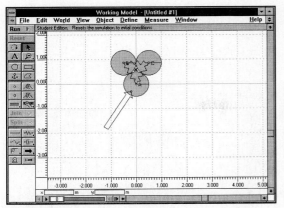

Figure 8.15

- Use a selection rectangle to **select all** the objects in the workspace.

- Open the **Appearance** window for the mixed selection.

- Click **off** all three **tracking** options (Track center of mass, Track connect, and Track outline). If you need help, refer to Tutorial 5.

Your screen should look like Figure 8.16.

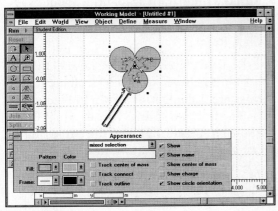

Figure 8.16

- Click on a **blank area** in the workspace.

- Select the **system center of mass** with the cursor.

Note that Working Model has given this point the name Point[10012], as shown in Figure 8.17. The system center of mass point will always be Point[10012].

Figure 8.17

- Click on **Track connect.**

- Close the **Appearance** window.

> **Tip** If you make a mistake and have unwanted tracking in your workspace, you can clean it up by selecting World menu, Erase Track.

Running the Simulation

You will now run the simulation until approximately $t = 1.3$ s so that you can observe the behavior of the simulation at specific times, and compare the results of Case 1 to those you achieve later in the tutorial.

- Open a **time** output meter with **digital** display.

- Run the simulation until about **t = 1.3 s.**

- Click **Reset.**

Use the Tape Player to examine the system behavior at specific times. At $t = 0.04$ s, the force is still acting as shown in Figure 8.18. The force becomes a dashed line at $t = 0.06$ s, as shown in Figure 8.19; this is when the force is no longer active.

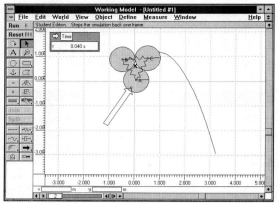

Figure 8.18

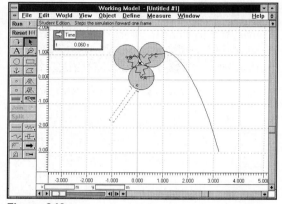

Figure 8.19

Observe that the path of the mass center is parabolic; this would be expected for projectile motion. This completes Case 1 of the given problem. (You may want to save your work at this point.)

Setting up Case 2

Now you will apply the spring constraints that are specified in Case 2 of the problem.

- Click **Reset.**
- Select springs **AB** and **BC.**
- Open the **Properties** window.

- Change the condition in the Active when field from Always to **t<.1**

Your screen should look like Figure 8.20.

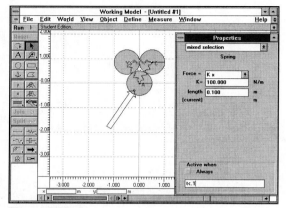

Figure 8.20

- Select spring **AC.**
- Change the condition in the Active when field from Always to **t<.6**
- Close the **Properties** window.
- **Run** the simulation again until about **t = 1.3 s.**
- Click **Reset.**

Use the Tape Player to observe how all the constraints behave. At $t = 0.08$ s, all the springs are still intact, as shown in Figure 8.21. The first two springs "break" at $t = 0.1$ s, as shown in Figure 8.22, and the last spring is intact at $t = 0.58$ s, as shown in Figure 8.23, and broken at $t = 0.6$ s, as shown in Figure 8.24. All of the spring failures occur at the times specified in Case 2 of the problem. Notice that, whether the springs break or not, the system mass center moves exactly the same way as in Case 1; compare the path of the mass center in Figures 8.19 and 8.24.

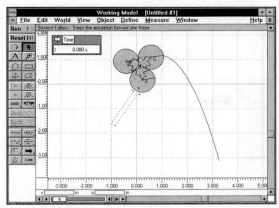

Figure 8.21

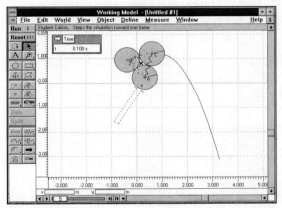

Figure 8.22

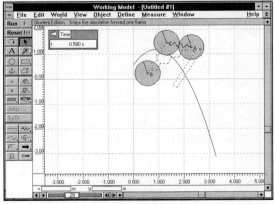

Figure 8.23

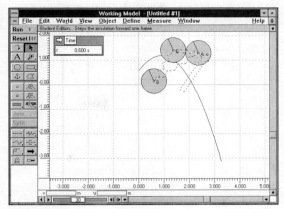

Figure 8.24

- Click **Reset.**

Creating New Reference Frames

The default *reference frame* for Working Model is the background, or world. Working Model also allows you to choose any object in the simulation as the current reference frame. When you select an object as the reference frame, the object remains stationary on the screen while other objects move relative to it. For example, in a model of the solar system, the sun is commonly used as the reference frame, and the planets rotate about it. If the earth is chosen as the reference object, the effect is similar to the pre-Copernican view of the solar system. The earth is viewed as a stationary object on the workspace, while other planets and the sun revolve around it. In Working Model, defining a frame of reference provides a "point of view" during the simulation and does not affect the coordinate values of any object in your model. All numerical measurements in Working Model remain the same, no matter what frame of reference you define.

There are two general uses for reference frames. You can use them to jump quickly between various views of a simulation (you can set Scroll and View individually for each reference frame). You can also use reference frames to watch a simulation from any object's point of view. You will create a new reference frame for this problem to view the system relative to its center of mass.

- Select the **system center of mass** (you may need to use the Properties window to select this point).

- Select **View** menu, **New Reference Frame...**

The New Reference Frame dialog box that opens shows the default name for the new reference frame, which in this case is System Center of Mass. Refer to Figure 8.25. You can change this name if you wish. You can also show an eye symbol and/or x, y axes at the center of the new reference frame. For this simulation, you will keep all of the defaults.

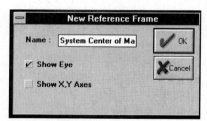

Figure 8.25

- Click **OK.**

The workspace now has an eye symbol at its system center of mass, as shown in Figure 8.26.

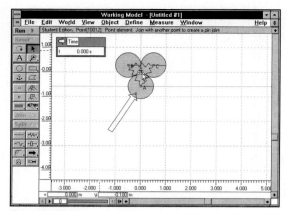

Figure 8.26

- **Run** the simulation again until about **t = 1.3 s.**

- Click **Reset.**

Now the only motion displayed is the motion relative to the system center of mass. Figure 8.27 shows the system at $t = 0.6$ s.

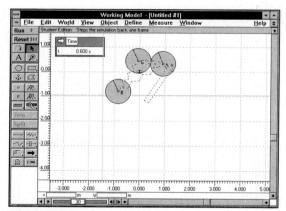

Figure 8.27

- Select **View** menu.

At the bottom of this menu all the reference frames are listed, as shown in Figure 8.28. You now have two reference frames listed: Home and System Center of Mass.

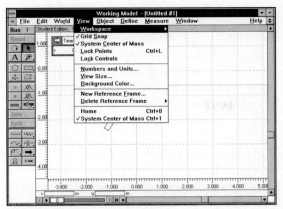

Figure 8.28

Note that there is a checkmark next to the current reference frame. You can select either of these reference frames on the View menu to switch back and forth between them, or you can use the keyboard commands listed on the menus to toggle between them.

- Click in the **workspace.**

W Type **Ctrl+0** to return to the world reference frame.

M Type **Command+1** to return to the world reference frame.

Obtaining Results

You will create a meter and customize its formula to display the total linear momentum of this system.

Customizing the Parameters in an Output Meter

When you create a meter, the equations that define the displayed output automatically appear in the Properties window for that meter. You can customize these equations, and the names of the variables they represent, to display output for any quantity that can be expressed with a formula. You will customize a meter for momentum to display the total linear momentum of this system.

- Open a **meter** for the **Momentum** of **mass A** in digital display.

- Open the **Properties** window for this meter.

- Maximize the size of the Properties window. If you need help, refer to Tutorial 2.

Your screen should look like Figure 8.29.

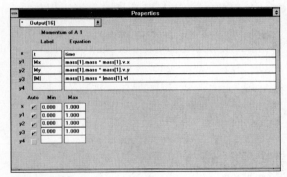

Figure 8.29

You will modify the equation for |M| on the fourth line under Equation. This equation represents the magnitude of the linear momentum of circle A, i.e., the mass of Mass[1] times the velocity of Mass[1]. (Circle A may have a different number in your case.) You will change this equation to represent the total system momentum, as required by the Problem Statement.

- Change the **label** on the graph from |M| to |**M**|tot on line **y3.**

- Change the **equation** on line y3 to: |**mass[1].mass*mass[1].v+mass[2].mass *mass[2].v+mass[3].mass*mass[3].v**|

Your dialog box should now look like Figure 8.30.

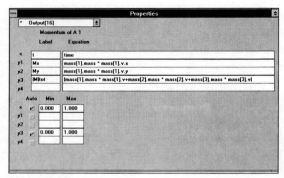

Figure 8.30

Note: To get the | for the equation, use Shift+\. It may look like a broken vertical line on your keyboard.

Tip You can use the usual copy and paste commands to copy and paste text in the dialog boxes and then modify the mass numbers, rather than retyping the entire equation.

Now the equation represents the total linear momentum of the three circles.

- Close the **Properties** window.

Notice that the title for the third line in the meter is now |M|tot.

- Use the **Appearance** window to change the title on the meter to **System Momentum.**
- Close the **Appearance** window.
- Click off the **Mx** and **My** boxes in the meter so only |M|tot will be displayed.

- **Run** the simulation until about **t = 0.6 s.**

The total linear momentum of the system is displayed in the meter. Figure 8.31 shows the value at $t = 0.6$ s. Note that the momentum varies with time, and so it is not conserved for this system; this is because there is an external force (actually an *impulse,* i.e., force acting over time) acting on the system, namely, the gravitational force from the earth.

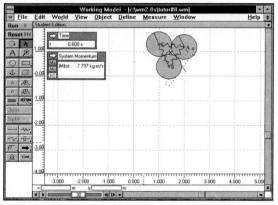

Figure 8.31

Link to Dynamics Determine the total kinetic energy of the system and the potential energy resulting from the height of the masses and the extension or compression of the springs, using dynamics equations and your simulation. Compare the results. Is energy conserved for this system?

- **Save** the simulation and **exit** Working Model.

8.1 The Basics Set up the workspace with grid, rulers, and axes. Create a 1-m circle centered at (-2,2) and connect a spring with a free length of 2 m between the circle's center and the workspace origin, as shown in Figure E8.1a. Make the spring active when the circle's mass center is below the origin and inactive when the circle's center is above the origin. (Hint: The circle's y position is written as mass[1].p.y.) Add a 2 N force to the circle oriented at a 45° angle. Track only the circle's center of mass and run the simulation. Add a second circle connected with a spring to the workspace origin as shown in Figure E8.1b. Show the system center of mass and set it as a new reference frame. Create a meter to show the relative horizontal velocity of the two masses and rerun the simulation.

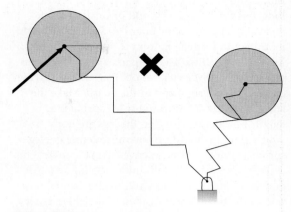

Figure E8.1b

8.2 Applying the Basics Model a 1,000,000-kg train with a velocity of 2 m/s heading toward the end of the track. It is approaching a 100,000-kg bumper. Turn off gravity, all friction, and make all elasticity zero. Examine the net horizontal force on the train first when the bumper is directly anchored to ground; reexamine it when the bumper is attached to the ground with a mammoth spring. Define the spring so that it is active only when the spring is compressing. (Use the variable constraint[•].dv.x for the velocity across the spring, where • is the proper object number for the spring, and dv is the differential, or relative, velocity between the ends of the spring.) See if you can design a spring constant to lower the maximum force on the train during impact. What are some other important design parameters? Save this simulation.

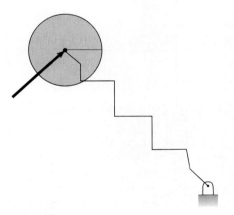

Figure E8.1a

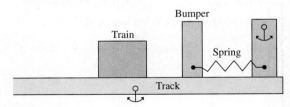

Figure E8.2

8.3 For Fun and Further Challenge

Summer is here and you are working on a new design for your diving board company using the model shown in Figure E8.3. The board is 6 m long, has a mass of 20 kg, and pivots on the left end. (Put a point on the board and the ground and join them to form a pin joint to serve as the pivot.) Model the diver with a 60-kg mass and a spring to represent the diver's leg action. Place the leg spring 0.5 m from the right end of the board. Make the leg stiffness 2000 N/m and the leg spring free length 1 m. Remember that the legs are not attached to the board, so you must define the leg spring so it is not always active. Start the diver 2 m above the board so he falls under the influence of gravity. Using the stiffness and the horizontal position of the board spring as design variables, see if you can maximize dive height.

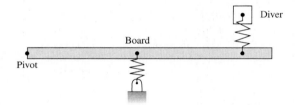

Figure E8.3

8.4 Design and Analysis
Create a ground with a small bump and design a vehicle to support fragile loads as shown in Figure E8.4. The wheels are 5 kg, the truck (bottom mass) is 20 kg, and the suspended mass is 200 kg. The springs are 10,000 N/m in stiffness. The applied force is 200 N and rotates with the mass. (The vehicle will rotate when it hits the bump.) Use gravity and air resistance. Measure and compare the vertical accelerations of the truck and suspended mass as the vehicle traverses the bump. What happens if you change the suspension stiffness?

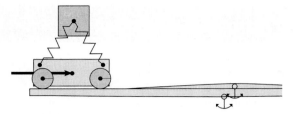

Figure E8.4

Exploring Your Homework   This problem includes a spring and a pulley in a mechanical system. Energy methods can help solve the problem analytically. What solution does Working Model give?

8.5
The system shown in Figure E8.5 is released from rest with the spring unstretched. If the spring constant is $k = 30$ lb/ft, what maximum velocity do the weights attain?

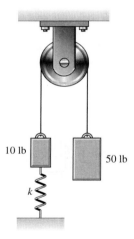

Figure E8.5

Exercise 8.5 is adapted from Bedford and Fowler's Engineering Mechanics: Dynamics *(Addison-Wesley, 1995), page 154.*

Ferris Wheel

Objectives

In this tutorial you will learn to:
- Duplicate objects
- Use the Motor tool
- Define motor properties
- Change the colors and patterns of objects
- Examine accuracy vs. speed for simulations
- Skip frames
- Export data from a simulation
- Use force fields
- Enter text in the workspace

Dynamics principles related to this tutorial

- Newton's laws for systems of particles
- Angular momentum for a system of particles
- Normal component of acceleration

In this tutorial you will look at a problem that demonstrates many dynamics principles: Newton's second law of motion for rigid bodies and systems of particles, angular momentum, and curvilinear components of acceleration. You will also learn how to export data from a Working Model simulation to a word processor or spreadsheet, how to set up a force field, and more about how Working Model's accuracy settings affect your results.

Problem Statement

The Ferris wheel shown in Figure 9.1 has 8 identical passenger buckets (A through H), dimensioned as shown. The entire assembly is made of steel and driven by a motor attached to the center of the wheel. Create a Working Model simulation for the Ferris wheel. If the motor is programmed to run at a speed of 2 rpm no matter what the load conditions are, export data to create a spreadsheet that will give the torque transmitted by the motor as a function of time for the following three cases:

1. When all the buckets are empty.

2. When the Ferris wheel is fully loaded; that is, every bucket contains two passengers, each weighing 125 lb.

3. When the Ferris wheel is fully loaded, as in Case 2, and a uniform wind force of 10 lb is blowing horizontally in the plane of the wheel (as shown in Figure 9.1).

Display vectors showing the force at the pin connection on bucket A and the acceleration of the mass center of bucket A. Display graphs showing the angular momentum of the wheel and the angular momentum of the mass center of bucket A as functions of time.

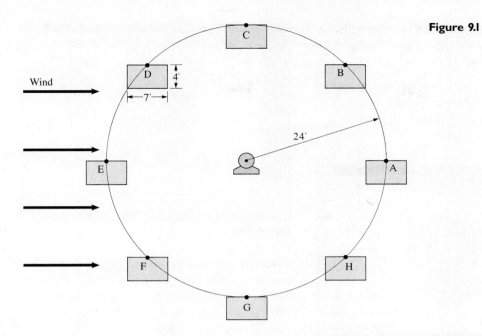

Figure 9.1

Setting Up the Workspace

- **Open** a new Working Model simulation with your workspace set up as before.

- Change the units to **English** (**slugs**); refer to Tutorial 7 if you need help.

- Be sure Accuracy is set to Accurate mode.

Creating the Components

- Create a **circle** of radius **24 ft**, centered at the origin.

Your entire workspace should be filled with the circle, as shown in Figure 9.2.

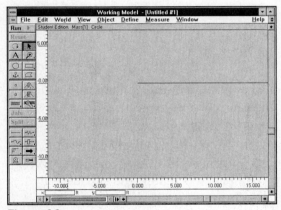

Figure 9.2

- Use the **Zoom Out** tool and the scroll bars to adjust the size of the workspace so that the entire object is visible.

- Use the **Properties** window to change the **material** of the circle to **Steel** and confirm that the **mass** is **92.117** slugs.

Your screen should look approximately like Figure 9.3. You will now create one bucket rectangle. You will duplicate it later to make the eight passenger buckets.

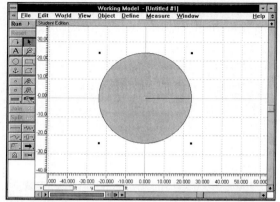

Figure 9.3

- Create a **rectangle 4 ft high** by **7 ft wide.**
- Change the **material** to **Steel** and confirm that the mass of the rectangle is **1.425** slugs.
- Use the **Point** tool to attach 8 points to the circle, approximately at the locations where the buckets will be attached, but well within the circle so that they won't accidentally attach themselves to the background.
- Attach one additional **point** to the rectangle.
- Remember to click on **Arrow** to deselect Point when you are finished.

Your workspace should look like Figure 9.4.

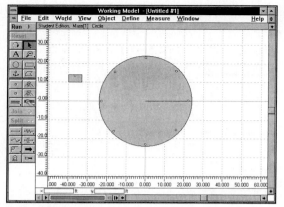

Figure 9.4

Since the eight points are on the x axis, on the y axis, or oriented at 45° from either axis, they will have local coordinate components equal to 24 ft or 24cos45 ft = 16.97 ft, in the directions indicated in Table 9.1:

	X (ft)	Y (ft)
A	24.000	0.000
B	16.970	16.970
C	0.000	24.000
D	-16.970	16.970
E	-24.000	0.000
F	-16.970	-16.970
G	0.000	-24.000
H	16.970	-16.970

Table 9.1

- Use the **Properties** window and Table 9.1 to move all the points to their proper locations.
- Use the local coordinates (**0,2**) to locate the point attached to the rectangle at the midpoint of the top edge.

Your workspace should look like Figure 9.5.

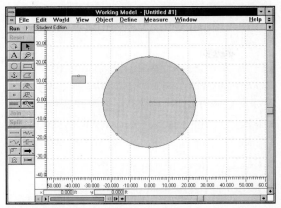

Figure 9.5

Duplicating Objects

To cut, copy, paste, and duplicate objects, use the corresponding commands on the Edit menu, just as you would in word processing or spreadsheet software programs, keeping in mind the following warning: objects that are being copied should not be constrained to other objects that are not being copied; otherwise a complicated constraint situation may result. In this case, you will duplicate the bucket *and its point* seven times *before* they have been attached to the Ferris wheel. The Duplicate option serves approximately the same function as Copy and Paste together.

- Select the bucket and point.

   Tip You can use the selection rectangle you learned in Tutorial 7 to conveniently select both objects.

- Select **Edit** menu, **Duplicate.**

Another bucket and point appear in the workspace; note that the new (duplicated) objects are now selected, as shown in Figure 9.6.

   Tip **W** Typing Ctrl+D will also duplicate selected items. **M** Typing Command+D will also duplicate selected items.

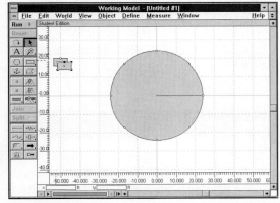

Figure 9.6

- **W** Type **Ctrl+D** six times.
- **M** Type **Command+D** six times.
- Temporarily **Anchor** the circle.
- Join one bucket to each point on the circle, using the **Join** button. Refer to Tutorial 7 for a review of the Join tool.

Remember that if you don't temporarily anchor the circle before using Join, the circle may move toward the bucket instead of the other way around. Clear the anchor when you are done.

- Use the Properties window and the coordinates in Table 9.1 to confirm that the location of the new constraints (the pin joints) did not shift during the join operation.

Your workspace should look like Figure 9.7.

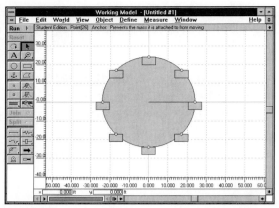

Figure 9.7

Using the Motor Tool

A motor creates a *rotational constraint* between two objects, or between one object and the background; it contains a built-in pin joint to attach it to the necessary object(s). Motors can exert torques necessary to constrain rotation, rotational velocities, and rotational accelerations. You can split rotational constraints, such as motors (using the Split tool), in order to edit their individual components, then rejoin them with the Join tool.

There are four types of motors:

1. Torque. Torque motors exert a specified twisting force between the two objects attached to the motor.

2. Rotation. Rotation motors exert whatever force is necessary to keep the objects attached to the motor oriented at a particular angle of rotation relative to each other.

3. Velocity. Velocity motors exert whatever force is necessary to keep the objects attached to the motor rotating at a constant angular velocity relative to each other.

4. Acceleration. Acceleration motors exert whatever force is necessary to keep the objects attached to the motor rotating at a constant angular acceleration relative to each other.

In this problem you will use a velocity motor to keep the Ferris wheel rotating at a constant angular velocity of 2 rpm (or 12 degrees/s).

- Click on the **Motor** icon on the tool bar (the bottom icon in the left-hand column).

- Click at the center of the circle.

A motor appears, as shown in Figure 9.8.

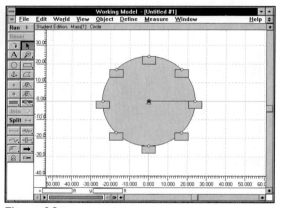

Figure 9.8

Defining motor properties

- Open the **Properties** window for the motor.

The Properties window displays a number of important details concerning the motor, as shown in Figure 9.9. It shows the name of the motor (Constraint[36] in this case), its type (Velocity is the default, but the Torque, Rotation, and Acceleration motors are available on the pull-down menu), its value (in units of °/sec), its base point (the background point, Point[26] in this case), its point (the point on the circle to which the motor is attached, Point[35] in this case), and the Active when box.

> Note: Your object numbers may differ from those in the figure, depending on the order in which you created your objects.

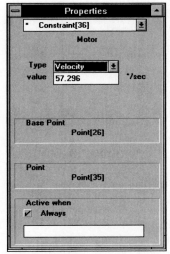

Figure 9.9

- Change the **value** to **12**°/sec.
- Verify that the points specified as the **base point** and the **point** are located at the origin of the workspace (i.e., the center of the circle).
- Close the **Properties** window.
- **Lock points** in their present locations.

Changing the colors and patterns of objects

You can use the Appearance window to change an object's fill pattern or color, or frame thickness or color. The frame is the object's outline.

> Note: If you are using a black and white printer, changing the colors will not change your printout, except for creating different levels of gray.

- Open the **Appearance** window.
- Label the bucket at 3 o'clock **A** and show its name.

In the lower left-hand corner of the Appearance window are the Fill Pattern and Color and Frame Pattern and Color options. Use the pull-down menus to examine all the options available, then modify the appearance of your Ferris wheel.

> Note: You may need to use the scroll bars to move your object above the Appearance window so that you can select all the buckets.

- Give all the buckets a thick frame. (Use the heaviest line width.)
- Select two different patterns or colors for the buckets, and alternate them.
- Select a pattern for the wheel.
- Close the **Appearance** window.

> *Tip* You can increase the color palette by re-selecting the color you're already using (the one with the dot on it, as shown in Figure 9.10 for Windows and Figure 9.11 for the Macintosh). A color dialog box pops up, as shown in Figure 9.12 for Windows and Figure 9.13 for the Macintosh.

W Figure 9.10

M Figure 9.11

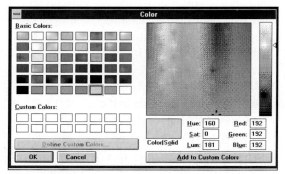

W Figure 9.12

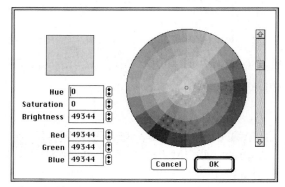

M Figure 9.13

Once you have changed the colors and patterns on your Ferris wheel, your workspace should look something like Figure 9.14.

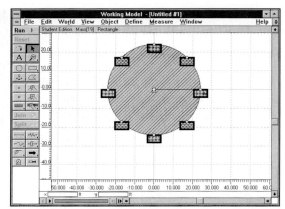

Figure 9.14

You will create four meters and two vectors to monitor your simulation.

- Create a **time** output meter in **digital** display.

- Create an output meter showing the **torque transmitted** by the motor in **digital** display.

- Create an output meter showing the **angular momentum** of bucket A in **graph** display.

- Create an output meter showing the **angular momentum** of the wheel in **graph** display.

- Create a vector showing the **total force** at the pin connection of bucket A.

- Create a vector showing the **acceleration** of the mass center of bucket A.

- Position the meters conveniently in the workspace.

   Note: You may need to use the scroll bars to move the x, y axes of the workspace so that they do not interfere with any of the meters.

Your workspace should look something like Figure 9.15. You won't see the vectors until you run the simulation. You can confirm that you have placed them, however, by using the

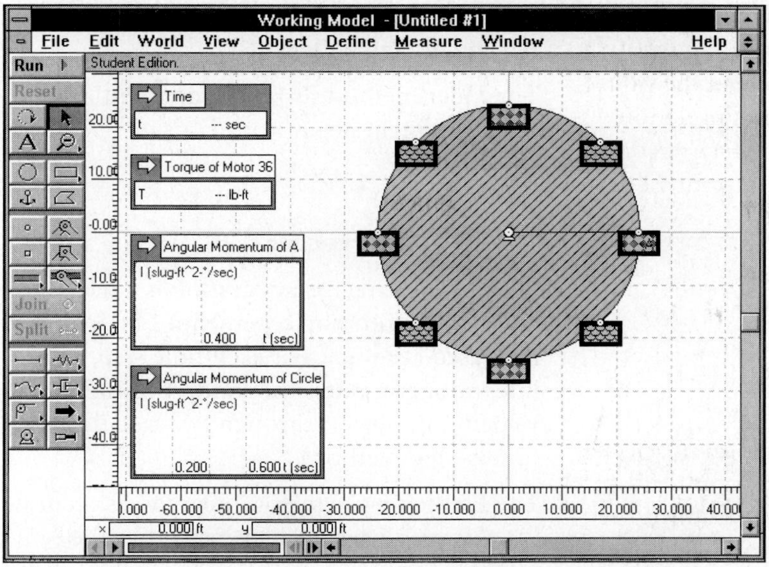

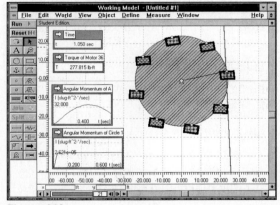

Figure 9.15

Define menu as follows.

- Select bucket A.
- Select **Define** menu, **Vectors.**

You should see a checkmark next to Acceleration, indicating that an acceleration vector is located on this object.

Running the Simulation

- **Run** the simulation until **t>1 s.**
- Click **Reset.**

Your vectors will look similar to those in Figure 9.16 as your simulation runs.

Figure 9.16

Accuracy and the Integrator

Notice that your simulation runs very slowly. As you learned in Tutorials 4 and 6, a simulation can be made to run faster if you are willing to give up some degree of accuracy in your results. To see if this simulation can be made to run faster without sacrificing too much accuracy, use the Tape Player to read the value of motor torque at $t = 1$ s for the simulation just run; it should be 373.019 lb-ft. (Depending on your system configuration, your values may vary slightly.)

- **Reset** your simulation.
- Select **World** menu, **Accuracy...**
- W Select **More.**
- M Select **More options.**

Working Model gives you the option of selecting from three different methods of integration for your simulation. The integrator is the mathematical process that continuously integrates objects' accelerations to update their positions and velocities. The following integrators are available in Working Model and appear to the left of your dialog box:

- Euler Integration
- Predictor-Corrector Integration
- Fourth-order Runge-Kutta (Runge-Kutta 4) Integration

In order to understand the relative complexity of the three methods, examine the following first-order differential equation:

$$x = f(x,t)$$

and see how each integrator solves it numerically. You want to solve x_{n+1}, or the value of x at the time of the "next" step, t_{n+1}, given the information x_n and t_n.

Euler

Euler integration is the fastest and simplest (but least accurate) integrator available for a given time step. Euler integration is the default in the Fast simulation mode and

should suffice to give you a rough idea of the motion of your model.

The Euler method solves the differential equation in a single step:

$$x_{n+1} = x_n + \Delta t \cdot f(x_n, t_n)$$

Predictor-Corrector

Predictor-Corrector integration is a relatively simple and common scheme for obtaining increased accuracy. For each time step, the integrator computes the states (position and velocity) of objects through two calculations (a two-step method).

The Predictor-Corrector integrator is accurate for common problems such as projectile motion, oscillators, fields, and rolling objects. Predictor-Corrector integration is the default integrator in Accurate mode.

The Predictor-Corrector method solves differential equations in a pair of computations, called *predictor* and *corrector* (hence the method's name):

predictor:

$$x_{n+1}^* = x_n + \Delta t \cdot f(x_n, t_n)$$

corrector:

$$x_{n+1} = x_n + \frac{\Delta t}{2}(f(x_{n+1}^*, t_{n+1}) + f(x_n, t_n))$$

where x_{n+1}^* is an intermediate variable used to compute x_{n+1}.

Runge-Kutta 4

This is the most accurate and stable integration method available in Working Model. If you select this integrator, the radio button next to Custom will automatically become highlighted. At each time step, the Runge-Kutta 4 integrator calculates the accelerations of each object, through four calculation steps. The method is considerably slower yet much more robust than the Euler integrator.

Runge-Kutta 4 integrates the differential equation as:

$$x_{n+1} = x_n + \Delta t \left[\frac{1}{6} f(x_n, t_n) + \frac{1}{3} f(x^*_{n+0.5}, t_{n+0.5}) \right.$$
$$\left. + \frac{1}{3} f(x^{**}_{n+0.5}, t_{n+0.5}) + \frac{1}{6} f(x^*_{n+1}, t_{n+1}) \right]$$

where

$$x^*_{n+0.5} = x_n + \frac{\Delta t}{2} f(x^*_n, t_n)$$

$$x^{**}_{n+0.5} = x_n + \frac{\Delta t}{2} f(x^*_{n+0.5}, t_{n+0.5})$$

$$x^*_{n+1} = x_n + \Delta t \cdot f(x^{**}_{n+0.5}, t_{n+0.5})$$

For the purposes of this tutorial, you will use the Fast mode, with its default Euler integration method.

- Click the radio button **Fast** at the left side of the dialog box.

The radio button to the left of Euler should also have become highlighted.

- Click **OK.**

- **Run** the simulation again until **t>7.5 s.**

Your simulation runs somewhat faster now. The value of motor torque at $t = 1$ s is now 432.977 lb-ft. (Your values may be slightly different.) The motor torque value of the Fast result is approximately 16% greater than the Accurate result. For many applications, this would not be an acceptable loss of accuracy; for the purposes of efficiency in this tutorial,

you will use the faster method of solution. See Appendix A for a detailed discussion of accuracy and the mathematical algorithms that are used in Working Model.

Skipping frames

Once a simulation has been run, and the results stored in the Tape Player, you can choose the Skip Frame option from the World menu to make it move even more quickly through its motion.

- Select **World** menu, **Skip Frames, 8 Step.**

- **Run** the simulation again.

The simulation now runs eight times faster, displaying every eighth frame, but only as long as it is repeating the portion of the simulation you have already run. You will note that it slows down again when it must begin calculating new frames of the simulation.

You want to export all the data from this simulation to a spreadsheet, not just that of every eighth frame, so you will now eliminate frame skipping.

- Click **Reset.**

- Select **World** menu, **Skip Frames, 1 Step.**

If your patterns are very dense, the output vectors may be difficult to view.

- If necessary, use the **Vector Display** dialog box to choose a thick line for your vector display.

- Use the Vector **Lengths** dialog box to change the vector length scales for the **acceleration** and **pin force** to **2.000** and **0.001**, respectively.

Your vectors should now look similar to those in Figure 9.17 as your simulation runs.

Figure 9.17

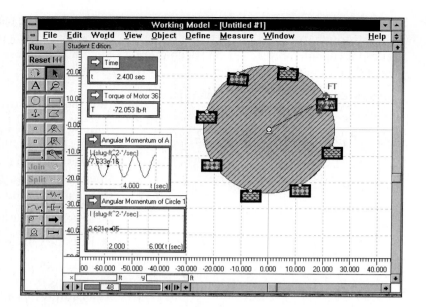

- Use **Save As** to save your simulation, with the tape recording, as **tutor#9a.wm**. For a review of Save tape recording, refer to Tutorial 6.

Exporting Data

There are two ways to export data from Working Model: (1) by exporting output meter data, and (2) by exporting a simulation's geometry data. In this tutorial, you will see how easy it is to export output meter data to a spreadsheet; the method you will learn will be similar for all types of data transfers. You will find additional information about data importing and exporting in Appendix C.

You can export meter data from Working Model in two ways:

1. Use the Export command to export meter data into a new spreadsheet or word processing file.

2. Use the Copy Data command to place the data from a meter into the clipboard; the data can then be pasted into a new or existing spreadsheet.

For this tutorial, you will first use the Export command to create a new spreadsheet for Case 1 (when all the buckets are empty); then you will use the Copy Data command to save the additional data for Cases 2 and 3 (full buckets, and full buckets plus wind force), which can be appended to the original spreadsheet.

- Select the **Motor Torque** output meter.

- Select **File** menu, **Export...**

The Export dialog box opens, as shown in Figure 9.18.

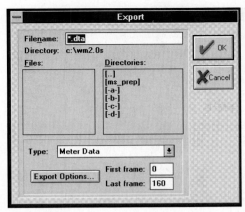

Figure 9.18

- Click the **Export Options...** button.

The Export Meter Data dialog box opens, as shown in Figure 9.19. There are a number of options available that you can use to control the data to be exported. The settings from this dialog box control the Copy Data command as well. Unless you change these options when you copy the data for Cases 2 and 3, they will remain in effect.

In this tutorial you will learn how to export your data to a Microsoft Excel spreadsheet. If you wish to export data to another spreadsheet or word processing program, you should make the appropriate changes.

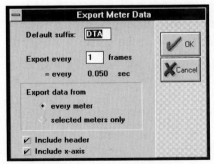

Figure 9.19

W Change **Default suffix** to **XLS** for a Microsoft Excel spreadsheet.

The next field in the Export Meter Data dialog box determines the interval at which frames will be exported. The default is every frame; if you export every frame in this simulation, the dialog box tells you that it will be at intervals of 0.05 s. You will leave this field at the default value for now. You will select the radio button labeled Export data from selected meters only. This will cause the new spreadsheet you are creating to output only the variables for whichever meter(s) you may select; otherwise Working Model will export data from all meters.

- Click on **selected meters only.**

Note that the Include header and Include x-axis options are switched on. Include header means the exported data will include information on the source file for your data. Include x-axis means that the exported data will include a column containing time (which is the x axis on the meter).

- Click **OK.**

You have returned to the Export dialog box (as in Figure 9.18). Notice that the bottom of the dialog box says:

> First frame: 0
>
> Last frame: 160

You may have a different number for Last frame, depending on how long you let your simulation run, as this value will match the last frame generated when you ran the simulation. The value in this box is the number of frames that will be exported or copied.

You will notice that changing the default suffix in the Export Meter Data dialog box also changed the file name in the Export dialog

box from *.dta, as shown in Figure 9.18, to
*.xls, which your screen should now show.

W In the **Filename** field at the top of the
dialog box, type **tutor#9.xls.**

W Click **OK.**

M In the **Export as** field, type **tutor#9.xls**.

M Click **Save.**

The simulation now runs until the last frame
is recorded and exported to Microsoft Excel.
The Exporting data... dialog box shown in
Figure 9.20 pops up on your screen while
Working Model is completing this export.

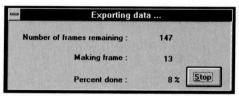

Figure 9.20

- Click **Reset** when the exporting has
  stopped.

W Return to the **Windows Program
Manager** to open **Microsoft Excel** and
the spreadsheet **tutor#9.xls.**

M Return to the **Finder** to open **Microsoft
Excel** and the spreadsheet **tutor#9.xls.**

Your spreadsheet should look like Figure 9.21.
(Your values may be slightly different.)

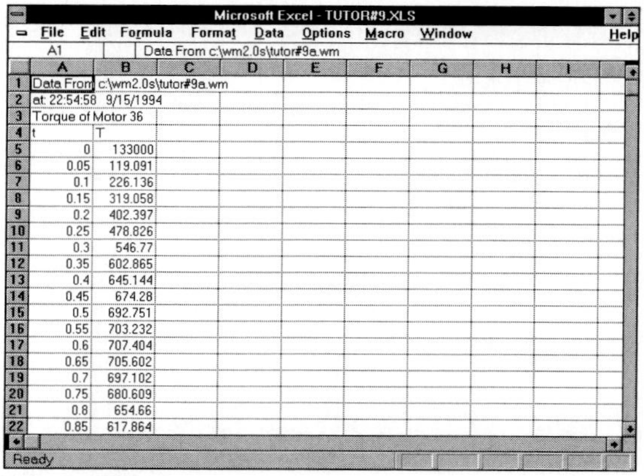

Microsoft Excel - TUTOR#9.XLS

	A	B	C	D	E	F	G	H	I
1	Data From	c:\wm2.0s\tutor#9a.wm							
2	at 22:54:58	9/15/1994							
3	Torque of Motor	36							
4	t	T							
5	0	133000							
6	0.05	119.091							
7	0.1	226.136							
8	0.15	319.058							
9	0.2	402.397							
10	0.25	478.826							
11	0.3	546.77							
12	0.35	602.865							
13	0.4	645.144							
14	0.45	674.28							
15	0.5	692.751							
16	0.55	703.232							
17	0.6	707.404							
18	0.65	705.602							
19	0.7	697.102							
20	0.75	680.609							
21	0.8	654.66							
22	0.85	617.864							

Figure 9.21

Tip Remember to check which directory
or folder your spreadsheet is being exported to; it should be your Working Model
directory or folder. You will need to open
that directory or folder to find your
spreadsheet.

Now you will run this simulation for Cases 2
and 3 of the Problem Statement.

- Return to Working Model.

- Select all the buckets and open the
  Properties window.

The mass for this mixed selection, i.e., the
mass of each bucket, is 1.425 slugs. You would
like to add 250 lb (250/32.2 = 7.764 slugs) to
each bucket, making the total mass 1.425 +
7.764 = 9.189 slugs for each bucket.

- Change the **mass** for all buckets to **9.189**
  slugs.

- Close the **Properties** window.

Now you will use the Copy Data command to export the data for this case to the clipboard (you cannot add data to an existing spreadsheet with the Export command—you can only create another new spreadsheet or replace the current one). To copy data, you must first store it in the Tape Player by running the simulation.

> **Tip** If you do not have enough memory to run Excel and Working Model at the same time, you can export the data for all three cases, then use Excel to copy the data from the separate files into the tutor#9.xls file.

- **Run** the simulation through one-quarter revolution.
- Click **Reset.**
- Use **Save As** to save your simulation and tape recording again (use a different name to preserve the data for Case 2).

The commands that you set before are still in effect, and Working Model will copy the data from the number of frames you have just run. You will make one change: you will eliminate the output of the time variable (you do not need a repeat column for time in the existing spreadsheet).

- Select **File** menu, **Export...**
- Select **Export Options...**
- Click off **Include x-axis** at the bottom of the dialog box.
- Click **OK.**
- Click **Cancel.**

This saves your Export options but doesn't export data to a spreadsheet file. A column for the time variable (the x axis of the output meter) will not be included in the data copied.

- Select the **Motor Torque** output meter.
- Select **Edit** menu, **Copy Data.**
- **W** Return to the **Program Manager** and open the **Clipboard Viewer** in the Main program group.
- **M** Return to the **Finder** and open **Edit** menu, **Show Clipboard.**

As shown in Figure 9.22, you have a listing of the torque at every 0.05 s until the point where you stopped your simulation. If you let it run slightly past 7.5 s, there will be a few more data points than those sent to the spreadsheet earlier using Export.

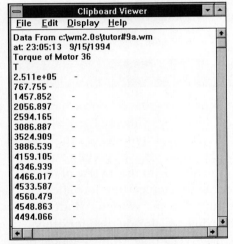

Figure 9.22

- Close the **Clipboard.**
- Open Microsoft Excel file **tutor#9.xls.**
- Select the top cell of the third column.
- Use the standard **Paste** command to insert the new data from the clipboard.

Your spreadsheet should look similar to Figure 9.23.

	A	B	C	D	E	F	G	H	I
1	Data From c:\wm2.0s		Data From c:\wm2.0s\tutor#9b.wm						
2	at: 22:54:58	9/15/199	at: 23:19:23 9/15/1994						
3	Torque of Motor 36		Torque of Motor 36						
4	t	T	T						
5	0	133000	251100						
6	0.05	119.091	767.755						
7	0.1	226.136	1457.852						
8	0.15	319.058	2056.897						
9	0.2	402.397	2594.165						
10	0.25	478.826	3086.887						
11	0.3	546.77	3524.909						
12	0.35	602.865	3886.539						
13	0.4	645.144	4159.105						
14	0.45	674.28	4346.939						
15	0.5	692.751	4466.017						
16	0.55	703.232	4533.587						
17	0.6	707.404	4560.479						
18	0.65	705.602	4548.863						
19	0.7	697.102	4494.066						
20	0.75	680.609	4387.739						
21	0.8	654.66	4220.453						
22	0.85	617.864	3983.234						

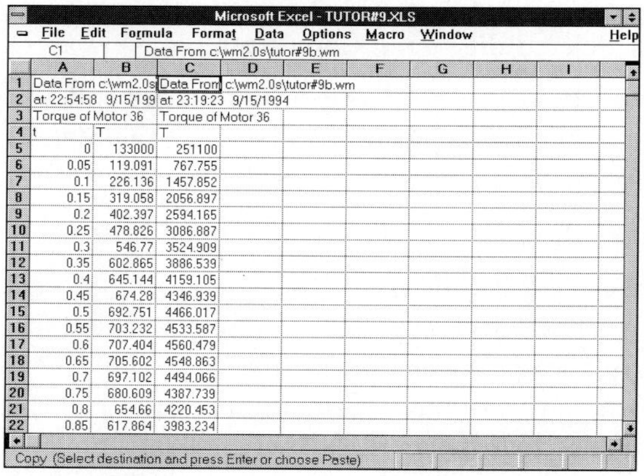

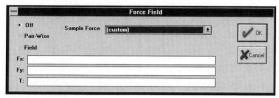

Figure 9.23

- Save **tutor#9.xls.**
- Return to Working Model.

Using Force Fields

You can define forces that act on each object or pairs of objects by using the Force Field command. You can also model unusual and sophisticated situations, e.g., gravity that grows proportionally to the inverse of distance. You can develop custom force fields using the same types of Working Model formulas that you have already used for other variables. In this example, you will model the 10-lb wind force by applying a horizontal force field to all objects.

- Select **World** menu, **Force Field...**

The Force Field dialog box appears on your screen, as shown in Figure 9.24. Open the pull-down menu to see the many force field options available. Some of these act on each object (e.g., Linear Earth Gravity), and some act on pairs of objects (e.g., Planetary Gravity). You can look at the governing formulas used for each of these, and refer to Appendix B to understand them better.

Figure 9.24

- Select **Wind** from the pull-down menu.

The default wind forces are fairly complicated functions of drag coefficient and cross-sectional area (refer to Appendix B). The wind force you will use is much simpler.

- Enter **10** for **Fx** (the horizontal force component), **0** for **Fy** (the vertical force component), and **0** for **T** (the torque applied by the force).
- Click **OK.**

Entering Text into the Workspace

The Text tool is used to enter text directly into the simulation workspace. You will use the Text tool to display the wind force as it is shown in Figure 9.25.

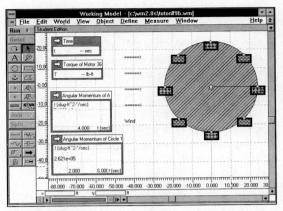

Figure 9.25

- Click on the **Text** icon on the tool bar (the letter **A**).

- Click in the workspace at approximately (**-45,15**).

- Type =====>.

- Press **Enter** three times.

- Repeat this arrow symbol three times; remember to hit **Enter** three times after each arrow symbol.

- Type **Wind.**

- Click on the **Arrow** icon to deactivate Text.

Text boxes behave just like other Working Model objects; you can click on them and move them, delete them, or change their size. You can choose from a wide variety of sizes, styles, and fonts for your text. To do so, select the text box that you want to change, then select **Object** menu, **Font.**

- **Run** the simulation through one-quarter rotation.

- Use **Save As** to save your simulation and tape recording (use a different name to preserve the simulation for Case 3).

- Select the **Motor Torque** output meter.

- Select **Edit** menu, **Copy Data.**

- Open the Microsoft Excel file **tutor#9.xls.**

- Select the top cell of the fourth column.

- Use the standard **Paste** command to insert the new data from the clipboard.

Your output data file is now complete.

	A	B	C	D	E	F	G	H	I
1	Data From c:\wm2.0s\	Data From	Data From	c:\wm2.0s\tutor#9c.wm					
2	at 22:54:58	9/15/199	at 23:19:2	at 23:16:55	9/15/1994				
3	Torque of Motor 36		Torque of	Torque of Motor 36					
4	t		T	T					
5	0	133000	251100	251100					
6	0.05	119.091	767.755	767.046					
7	0.1	226.136	1457.852	1457.941					
8	0.15	319.058	2056.897	2058.585					
9	0.2	402.397	2594.165	2597.9					
10	0.25	478.826	3086.887	3093.4					
11	0.3	546.77	3524.909	3535.337					
12	0.35	602.865	3886.539	3902.061					
13	0.4	645.144	4159.105	4180.509					
14	0.45	674.28	4346.939	4374.449					
15	0.5	692.751	4466.017	4499.364					
16	0.55	703.232	4533.587	4572.105					
17	0.6	707.404	4560.479	4603.112					
18	0.65	705.602	4548.863	4594.108					
19	0.7	697.102	4494.066	4539.964					
20	0.75	680.609	4387.739	4431.976					
21	0.8	654.66	4220.453	4260.538					
22	0.85	617.864	3983.234	4016.711					

Figure 9.26

If you examine the spreadsheet (and Figure 9.26), you will find that the required torque is considerably higher for the fully loaded Ferris wheel, and slightly more when the wind load is acting.

The angular momentum output meters show that the angular momentum for the wheel is conserved; you would expect this, because angular momentum is the product of radial distance, mass, and velocity, all of which are held constant for the wheel.

The vectors that show the force at the pin connection, and the acceleration of the mass center, illustrate a wide range of fluctuation for the three different cases as the Ferris wheel rotates.

Practice some of the skills you have learned here and explore the behavior of the Ferris wheel further.

1. Compare the spreadsheet results already obtained with the case where only two consecutive buckets are fully loaded, and the rest are empty (include the wind force).

2. Write the Newton's law equations of motion that will yield the pin force and linear and angular acceleration for one of the buckets. (Working Model can be used to display these values through output meters.) Note that in your simulation, the acceleration of the bucket is usually acting in the general direction of the center of the wheel. Acceleration directed towards the center of the path of motion is a normal component resulting from the rotation of the wheel; any component of acceleration which is not acting towards the wheel center would be due to the motion of the bucket relative to the wheel.

3. Write an equation for the angular momentum of one of the buckets. The buckets are affected by gravity and they behave like pendulums; therefore their angular momentum is not conserved.

4. Write an equation for the total angular momentum of the wheel/bucket system.

- Save **tutor#9.xls** and exit Microsoft Excel.
- Save **tutor#9.wm** and exit Working Model.

9.1 The Basics With your Ferris wheel experience behind you, you are ready for the Tea Cups (Fantasyland, Disneyland, not far from Dumbo). Actually, this is not quite as complicated as the real thing. Build a 50-ft diameter platform with two 15-ft cups arranged symmetrically (duplicate the first cup to make the second). Change the patterns and colors of the system's appearance for a pleasing effect. Connect a motor between each cup's center and the platform, and between the platform center and the background. (Selecting the motor tool and clicking in the desired locations will do this.) Place a point A on the edge of one of the cups to serve as a reference location. Calculate and set the torque of the motors to achieve the acceleration you desire. (Hint: Use a simple dynamic calculation based on the mass moment of inertia of the platform found in the Properties window.) The platform motor should have about ten times as much torque as the cup motors. Create meters for position and velocity for point A. Run the simulation using Fast mode, then switch to Accurate mode and examine the difference in the output at $t = 2$ s. Add a force field that represents wind and see if it has any impact on the system. Export data from the meters to a spreadsheet and graph the data in the spreadsheet.

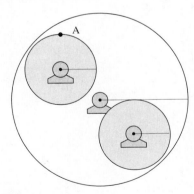

Figure E9.1

9.2 Applying the Basics You are working on your Airframe and Powerplant (A&P) license by examining the dynamics of a reciprocating aircraft engine. Create a 500-slug engine and a 30-slug prop as shown in Figure E9.2. Position their centers coincidentally and connect them with a torque motor of 200 ft-lb. Turn off gravity and run the simulation monitoring prop speed and any other outputs of interest. Can you calculate the prop speed after 10 seconds and compare it to the Working Model result? What happens to the engine during the simulation and how would you address this problem as an aircraft designer? Monitor prop, engine, and total system angular momentum and explain the results.

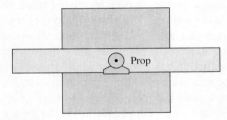

Figure E9.2

9.3 For Fun and Further Challenge You desperately need a low-cost automated sanding device for your business and you have come up with the concept illustrated in Figure E9.3, where a pulley system attached to the background connects the motor/wheel assembly to a 10-slug block. A spring connects the block and the background. The wheel rotates at 1 rps and the spring should be created so it keeps the pulley rope tight. Assuming a friction coefficient of 0.8, find the minimum stiffness and free length for the spring necessary for smooth operation of the sanding block.

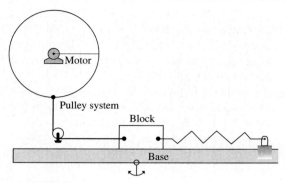

Figure E9.3

9.4 Design and Analysis Design a four-wheel drive vehicle moving down a smooth road, as shown in Figure E9.4. First add constant magnitude torque motors to each wheel and create controls in the workspace to allow you to vary the torque as you examine front and rear wheel drive dynamics. Set friction coefficients to 1 and elasticity to zero. Can you figure out what torque is necessary to accelerate from 0 to 60 mph in 10 seconds? Can you make the vehicle "pop a wheelie"? Can you explain why it happens? Change the motor torque to be a function of the wheel's rotational velocity, so the motors have maximum torque at zero velocity and less torque as the wheel's rotational velocity increases. (This is more realistic for motors and engines.) See how these new motors perform. What happens if you add wind to the simulation?

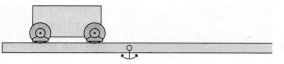

Figure E9.4

Exploring Your Homework This problem involves a constant velocity motor and a slider-crank linkage.

9.5 The crank *AB,* shown in Figure E9.5, has a constant clockwise angular velocity of 200 rpm. What are the velocity and acceleration of the piston *P*? Hint: Use a length actuator for link *BP.* Fraction and elasticity in this problem are negligible.

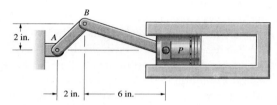

Figure E9.5

Exercise 9.5 is adapted from Bedford and Fowler's Engineering Mechanics: Dynamics *(Addison-Wesley, 1995), page 268.*

Design of a Vibration Absorber

Objectives

In this tutorial you will learn to:
• Use global coordinates for locating points
• Click and drag to apply an initial displacement
• Use Pause Control
• Create dampers and damped springs
• Attach pictures to objects

Dynamics principles related to this tutorial

• One- and two-degree-of-freedom vibrating systems
• Resonance
• Free vibrations and damped vibrations
• Forced vibrations

Introduction

Many engineering problems are concerned with vibrations. Usually, the engineer wishes to reduce the intensity of oscillating motion in a system. Large structures, for example buildings, bridges, or airplane wings, can fail as a result of undesirable resonant vibrations. In this tutorial, you will explore spring-mass-damper systems to see how Working Model can be used for vibrations problems. You will also see how to copy pictures from drawing applications and attach them to your simulations.

Problem Statement

The frictionless system shown in Figure 10.1(a) consists of mass m_1 and spring k_1; this system represents a machine which contains an unbalanced motor and is attached to a wall. The imbalance of the motor has a frequency equal to the natural frequency of the spring-mass system. A resonance situation results. This problem can be overcome in two ways:

1. The spring constant k_1 can be varied so that the natural frequency of the spring-mass system is changed, and resonance no longer occurs.

2. For cases in which k_1 cannot be altered, another approach to vibration absorption must be taken. A secondary spring-mass-damper system can be attached to the original system, as shown in Figure 10.1(b). This arrangement is known as a *vibration absorber*; it will change the natural frequency of the entire system to prevent resonance.

You will create a simulation that demonstrates both of these methods. Assume that $m_1 = 1$ kg, $k_1 = 50$ N/m, $m_2 = 0.25$ kg, and the initial dimensions are as shown in Figure 10.1. Attach a picture of an unbalanced motor to m_1.

Figure 10.1

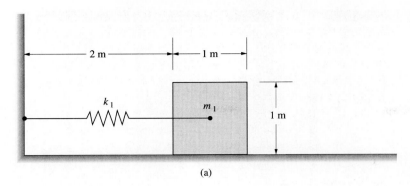

(a)

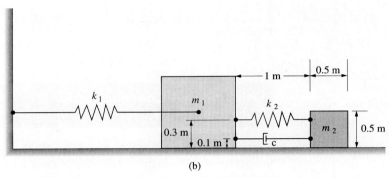

(b)

Starting Working Model

- **Open** a new Working Model simulation, with the default workspace you have used in each tutorial. The unit system and the accuracy mode should be the Working Model defaults, SI(degrees) and Accurate.

Creating the Components

You will use the Polygon tool to create an L-shaped polygon representing the wall and the ground on which the system will oscillate.

Placing objects will be more convenient if the inner corner of the L-shape is at the origin of the axis system.

- Create an L-shaped **polygon**: the left side of the polygon should be **1** m wide by **4** m tall; the bottom should be **8** m wide by **1** m tall; its inner corner is located at the **origin**.

- Scroll the workspace so the x, y axes are near the lower-left corner.

Refer to Tutorial 7 to review creating polygons. Your wall/ground polygon should resemble the shape in Figure 10.2.

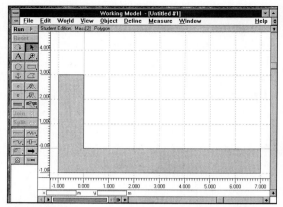

Figure 10.2

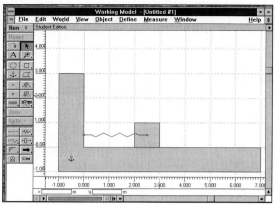

Figure 10.3

- **Anchor** the polygon.

- Create a **1-m square** with its center at
  (**2.5,0.5**) and its bottom edge on the
  ground.

- Confirm that the mass of the square is equal
  to **1 kg**.

- Label the square **m1**.

- Attach a horizontal **spring** to the wall and
  the center of mass m1. Remember to locate
  the endpoint of the spring in a bit from the
  edge of the wall so that it attaches to the
  mass and not the background.

Refer to Tutorial 8 to review attaching springs.
Later, you will correct the positions of the
endpoints of the springs in the Properties
window. Your workspace should look similar
to Figure 10.3.

Using Global Coordinates for Locating Points

When a point is attached to an object, its
Properties window shows two sets of coordinates: local or relative (relative to the mass to
which the point is attached), given at the top
of the Properties window; and global (relative
to the world axis system), labeled at the bottom of the Properties window, as shown in
Figure 10.4. You can set the values of either,
whichever is more convenient. You want to be
sure that the two spring points are located
properly; it will be easier to use the global
coordinates for the point that is attached to
the wall, Point[5], and the relative or local
coordinates for the endpoint attached to m_1,
Point[6]. (Depending on the direction in
which you created your spring, your point
numbers may be reversed.)

Note: When a point is attached to the
background, its local and global coordinates
are the same.

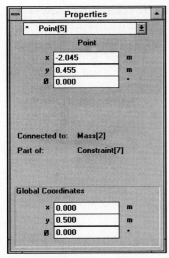

Figure 10.4

- Use the **Properties** window to locate the left spring endpoint at **global** coordinates (**0.0,0.5**); make sure that the point is attached to the wall.

- Locate the right spring endpoint at the **relative** coordinates (**0.0,0.0**) of m1.

Now you will make the surfaces of m_1 and the wall frictionless, then add the meters and controls needed for the simulation. Refer to Tutorial 3 for a review of creating controls, if necessary.

> **Tip** To select an item that is already part of a multiple-item selection, you must click away from all objects in the workspace and then select the item, or hold the Shift key down while you click to deselect the unwanted item(s).

- Select the wall and m1.

- Set *both* **friction** coefficients to **0.0** for the wall *and* m1.

- Open a **time** output **meter** in **digital** display.

- Open a **meter** showing the **x graph position** of m1 in **graph** display.

- Select the spring and open a **control** for the **spring constant**.

- Name and position the meters and the control appropriately, as shown in Figure 10.5.

Figure 10.5

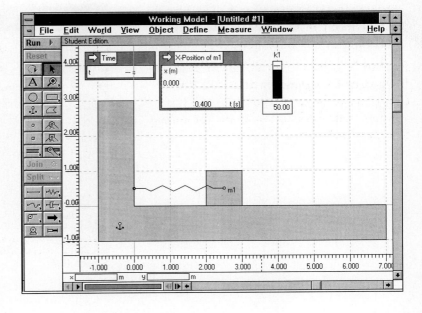

- Use the Properties window to check to make sure that the spring length is the same as its current length, **2.5** m.

- Set your control for the spring constant to **50** N/m if necessary.

For a review of working with the spring length, refer to Tutorial 8.

Your workspace should look like Figure 10.5. Because your spring constraint is so close to the edge of the wall mass, you may see a broken line from its endpoint to the wall's mass center. This will not affect your results.

Clicking and dragging to apply an initial displacement

Just as you can click and drag on meters and other objects to move them in the workspace, you can click and drag on the mass m_1. The difference in this case is that m_1 is constrained by the spring, so clicking and dragging on it stretches or compresses the spring. This is a feature of Working Model's Smart Editor, which maintains the relationships among elements in your simulation and allows you to edit the mechanisms you create without redefining all the constraints. You will learn more about the Smart Editor in Tutorial 11.

- Move **m1** horizontally to the left **1** m.

Refer to Figure 10.6.

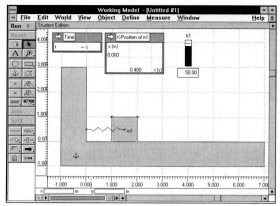

Figure 10.6

Look in the spring's Properties window, as shown in Figure 10.7. The free length of the spring is still 2.5 m, but now its current length is 1.5 m. The spring has been compressed 1 m.

If your Properties window doesn't show these values, change them now to match those in Figure 10.7. Before you leave the Properties window, confirm that mass m_1 is located at coordinates (1.5,0.5).

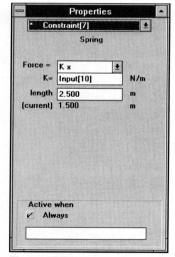

Figure 10.7

Using Pause Control

The Pause feature enables you to automatically stop a simulation when some condition is met. In this case, you will stop and reset the simulation when *time* > 6 s.

- Select **World** menu, **Pause Control...**

The Pause When dialog box appears on your screen, as shown in Figure 10.8.

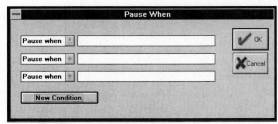

Figure 10.8

W Click the **New Condition** button at the bottom of the dialog box.

M Click the **New Condition** button at the top of the dialog box.

The sample formula, *time* > 1.0, appears as the first pause condition (you can set up to three conditions).

- Click on the top pull-down menu, which currently says **Pause when**.

A menu of options appears that you can use to control your simulation. In addition to pause, you can set conditions that will cause the simulation to stop, reset, and loop. Use the delete option to remove conditions you no longer want.

- Select **Reset when** from the pull-down menu.

- Change the formula to read **time > 6.0**

- Click **OK**.

Running the Simulation

- **Run** the simulation until it stops and resets itself.

The simulation stops and resets itself at *time* = 6 s. Your workspace should look like Figure 10.9.

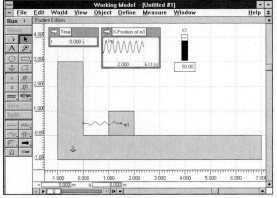

Figure 10.9

In order to make this spring-mass system resonate and model the given problem, you need to force the system to oscillate at its natural frequency. You can do this by applying a forced vibration to the system.

- Use the **Tape Player** to move the simulation to where it has completed one cycle of motion.

A dot appears on the x-position output graph, as shown in Figure 10.10. The time meter reads 0.88 s; this is the approximate period for one cycle of motion and will be used to calculate the approximate natural frequency of the system. A period of 0.88 s is equal to a frequency of $\omega_{approx} = 2\neq/0.88$ or 7.14 rad/s. Note that the exact natural frequency of the system could be obtained by using the formula

Figure 10.10

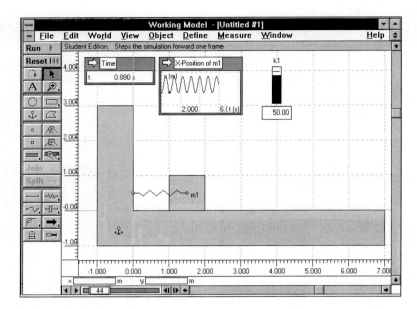

$$\omega_{exact} = \sqrt{\frac{k_1}{m_1}} = \sqrt{\frac{50}{1}} = 7.071 \text{ rad/s}$$

Resonance occurs when the frequency of the forced vibration is the same as the natural frequency of the system on which it acts.

You will now apply a forced vibration in the form of a *forcing function*, $F_x = F\sin\omega t$. Since the natural frequencies (ω_{approx} and ω_{exact}) have been calculated in units of rad/s, you will change the Working Model angle measure unit to radians; this will make it easy to use the calculated frequencies in your equations.

- **Reset** the simulation.

- Change the units to **SI (radians)**.

- Move **m1** back to its original position, **(2.5,0.5)**.

- Apply a **force** to the mass center of m1, horizontally from the left. (It is easier to click and drag away from the mass center to locate the force so you don't accidentally move the spring.)

- Use the force's Properties window to identify the base point.

- Use the base point's Properties window to move the base point to **relative** coordinates **(0,0)** on m1.

You will now define the forcing function to have the estimated natural frequency, ω_{approx}.

- Use the force's Properties window to set **Fx = 5*cos(7.14*t)** and **Fy = 0**.

- **Run** the simulation again.

When the simulation is finished running and has reset itself, your workspace should look like Figure 10.11. The system oscillations grow over time, an indication that the system is close to resonance.

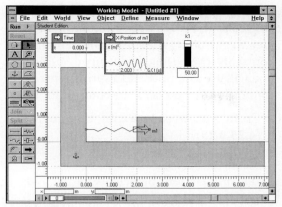

Figure 10.11

Now change the simulation to run at the exact resonant frequency, ω_{exact}.

- Use the force's Properties window to change **Fx** to **5*cos(7.071*t)**, using the exact resonant frequency, 7.071, for the forcing frequency.

- **Run** the simulation again.

Your X-position graph won't look dramatically different, but you will see that the oscillations grow large more quickly. You could confirm this by comparing the results obtained with the exact forcing frequency to those obtained with the approximation as follows: You could use the Tape Player to step out to the highest point on the graph for a given cycle, say the fifth; you could then switch from the output meter graph display to the digital display of the x position and get a measurement of the amplitude for that cycle. After running the simulation again with the first frequency, you could compare the amplitude for the fifth cycle. You would find that the simulation using the exact resonant frequency reaches a greater amplitude faster.

Obtaining Results

Now that your system has been set up and resonance achieved, you are ready to begin testing methods for avoiding resonance. Changing the value of the spring constant is the first method of avoiding resonance suggested by the Problem Statement. You will leave the forcing frequency at $\omega_{exact} = 7.071$, and change the value of the spring constant k_1.

- **Save** your simulation as **Tutor#10**

- Using the **k1 control**, change the spring constant to **25.25** N/m.

- **Run** the simulation again.

When the simulation resets, your workspace should look like Figure 10.12. Because the new natural frequency of the system,

$$\sqrt{\frac{k_1}{m_1}} = \sqrt{\frac{25.25}{1}} = 5.025 \text{ rad/s}$$

is now different from the excitation frequency (7.071 rad/s), resonance no longer occurs; the output is now a combination of the new natural frequency (if there were any damping in the system, this transient effect would eventually die out) and the forcing frequency (in the presence of damping, this "steady state" effect would dominate).

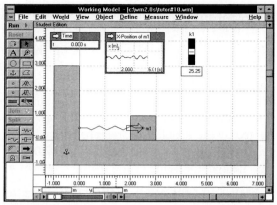

Figure 10.12

You will now return k_1 to its original value and create a secondary vibrating system, which will act as a vibration absorber. This is the other method of avoiding resonance that the Problem Statement suggests.

Creating Dampers and Damped Springs

Dampers, sometimes called dashpots, exert a force proportional to the difference in the velocity of their endpoints. The Damper tool is located below the Spring tool on the tool bar. Dampers are similar to shock absorbers in a car, or pistons that keep doors from slamming. The greater the *damper constant,* the better the damper resists high-velocity movements. You can change the damper constant in the Properties window (and you can create nonlinear dampers by using formulas).

The vibration absorber that you will create includes a spring as well as a damper. Both are attached to the primary spring-mass system. You could use the Damped Spring tool (the third option on the Spring icon on the tool bar). This tool permits you to assign values to

both the spring and damping constants in a single Properties window. However, you can't create controls for the spring and damping constants when using the Damped Spring tool. You will use a damper and a spring separately for now, and create separate controls for them.

- Create a **0.5-m square** with its center at (**4.25,0.25**).

- Label the square **m2**.

- Set *both* **friction** coefficients for m2 to **0**.

- Be sure the **mass** of m2 is equal to **0.25 kg**.

- Click on the **Damper** icon on the tool bar (below the Spring).

- Attach the damper between m1 and m2 at about **y = 0.1** m in the same way you would attach a spring.

- Create a **spring** between the two masses above the damper at about **y = 0.3** m.

- Use the **global** coordinates in the Properties window to locate the y positions of the spring endpoints at **y = 0.3** and the damper endpoints at **y = 0.1**.

- Use the **relative** coordinates in the Properties window to place the x positions of the spring and damper endpoints on m1 at **x = 0.5**, and the spring and damper endpoints on m2 at **x = -0.25**.

- Make sure the current length and the free length of the spring are both equal to **1** before continuing.

- Create **controls** for the new **spring** and **damper constants**.

Note: In the Properties window, the damper constant is called k; in this tutorial, we refer to the damper constant with the variable c. The second spring constant is k_2.

- Name and position the controls appropriately, as shown in Figure 10.13.

- Change **k1** back to **50** N/m.

- Set **k2** to **50** N/m and **c** to **1** if necessary.

- Change the **pause control** to let the simulation run for 12 s.

Your workspace should look like Figure 10.13.

- **Run** the simulation again.

At $t = 11$ s, your workspace should look like Figure 10.14. So far the vibration absorber is not working very well; the system is no longer resonating, but it still has considerable oscillations, as shown in the x-position output meter in Figure 10.14.

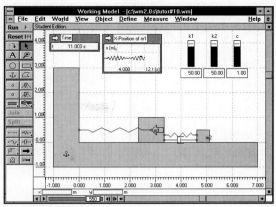

Figure 10.14

The equations governing the motion of this system, i.e., equations describing forced vibrations with damping of a two-degree-of-freedom vibrating system, can be found in any basic vibrations textbook. The variables for the present problem, m_1, m_2, k_1, the resonant frequency, and the applied driving force, can be used in these equations to estimate new

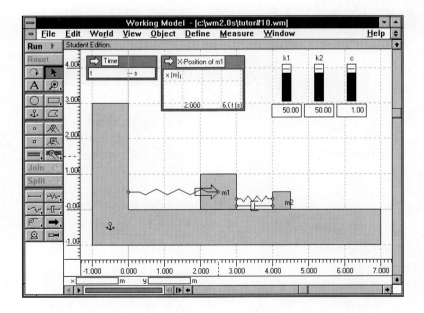

Figure 10.13

values for k_2 and c which will cause the oscillations to be better absorbed. For this tutorial, these values have been calculated for you.

- Using the **controls**, change **k2** to **10.4** N/m and **c** to **0.25** N-s/m.

- **Run** the simulation again.

After the simulation runs and resets, your workspace should look like Figure 10.15. The system vibrations were substantially reduced, and resonance is no longer a problem.

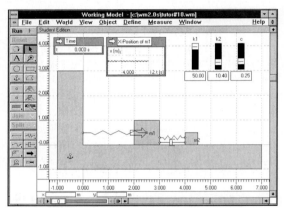

Figure 10.15

You can create an equivalent system using the *Damped Spring* tool.

- **Clear** the **spring** and **damper** between m1 and m2.

- **Clear** the **controls** for **k2** and **c**.

- Select the **Damped Spring** icon from the Spring options on the tool bar (the far-right option when you click and hold on the Spring icon).

- Connect the damped spring to m1 and m2 at **global** coordinates (**3,0.25**) and (**4,0.25**), respectively.

- Use the Properties window to change the **spring constant** to **10.4** N/m and the

damper constant to **0.25** N-s/m; make sure the damped spring length is equal to the current length, **1**.

- **Run** the simulation again.

The system behaves precisely as it did when you last ran it. Your workspace should look like Figure 10.16.

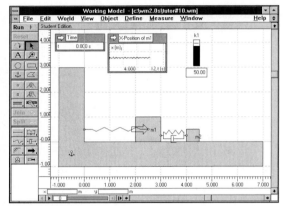

Figure 10.16

- **Save** your simulation.

Attaching Pictures

Picture objects are created whenever graphics data is pasted into the workspace. Working Model for Windows accepts metafile data, and Working Model for the Macintosh accepts PICT data. You can drag, cut, copy, and paste picture objects. You can also attach picture objects to a mass, but picture objects do not zoom or rotate, so it is best to attach them to items that will only translate. You can use any paint or drawing program available for creating pictures, as long as they can be handled as metafile or PICT images. The example below uses **W** Paintbrush or **M** MacPaint. Use Figure 10.17 as a guide for drawing the schematic of an unbalanced motor. If you do not have a drawing application, you can continue the tutorial without attaching a picture.

M If you do not have a drawing application, you can attach a random picture from the Scrapbook.

- **Minimize** Working Model.

W Use **Paintbrush** to create a schematic of an unbalanced motor.

W While in Paintbrush, **select** and **copy** the drawing.

M Use **MacPaint** to create a schematic of an unbalanced motor.

M While in MacPaint, **select** and **copy** the drawing.

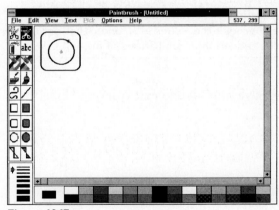

Figure 10.17

- Return to your Working Model simulation.
- Select **Edit** menu, **Paste**.

The picture appears as an object in your workspace, as shown in Figure 10.18. You can

change its size, if necessary, the same way you size other items. You will now attach it to the mass m_1.

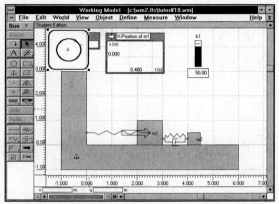

Figure 10.18

- Select **m1**.
- Hold down the **Shift** key and select the picture.
- Select **Object** menu, **Attach Picture**.

The picture is now attached to m_1, and it will move with mass m_1 when you run the simulation. Your screen should look like Figure 10.19. You can detach the picture by selecting m_1 and then selecting Detach Picture from the Object menu.

- **Run** the simulation.

If your picture happens to move behind the mass, use the Properties window to select it (it is labeled Picture), and then use Object menu, Move To Front.

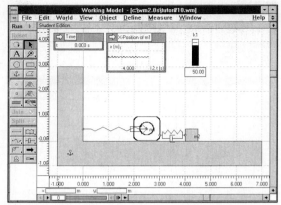

Figure 10.19

You have examined two ways to avoid resonance. A third way would be to just add a damper to the original system, Figure 10.1(a). However, sometimes the original system cannot be altered at all; that is when you would want to use a vibration absorber.

Try adding a damper to the original system to prevent resonance.

Try different values for k_1 from the original system and different values for k_2 and c from the vibration absorber.

Link to Dynamics

The effect of the vibration absorber shown in Figure 10.1(b) can be optimized by using the equations for a two-degree-of-freedom vibrational system. This optimization is not a trivial exercise, since there are so many variables, including the applied load and its forcing frequency, the frequency of the initial spring-mass system,

$$\sqrt{\frac{k_1}{m_1}}$$

the frequency of the absorbing system,

$$\sqrt{\frac{k_2}{m_2}}$$

the ratio of the absorber mass to the main mass, m_1/m_2, the ratio of the frequency of the applied load to the frequency of the initial spring-mass system, etc. Look up these equations in a vibrations textbook, and see if you can predict the values of k_2 and c that optimize the reduction in oscillations of m_1.

- **Save** your file and **exit** Working Model.

10.1 The Basics Create an 8 m platform anchored to the background, as shown in Figure E10.1. Add two 2-slug, 1-ft diameter wheels, both connected to the ground with springs with stiffnesses of 10 lb/ft and 3 lb/ft. Use global coordinates to help place the springs' ends all at 0.5 ft above the platform and at the centers of the wheels. Add a damper between the wheel centers with a coefficient of 2 lb-sec/ft. Drag one of the wheels about 1 ft horizontally as an initial condition and run the simulation, graphing wheel displacements and spring forces. Add a pause control to pause when time is greater than 1 s. Can you figure out how to pause when the left wheel changes direction? Draw a picture of a spoked wheel in another application, cut and paste it into Working Model, and attach it to one of the wheels. Rerun the simulation and examine the motion of the spoked wheel.

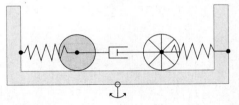

Figure E10.1

10.2 Applying the Basics Since the housing market is booming, you are starting a new venture, Don's Door Design. You are working on the automatic closing mechanism and need to design the spring and damper characteristics. Model the door as a 3-ft wide, 20-lb rectangle hinged at one end (use a pin joint that will attach the door to the background). Attach a spring and damper between the other end of the door and the background, as shown in Figure E10.2. Your job is to estimate the applied force a person should need to apply to the door and how long it needs to be applied (use the Active when field). Then pick spring and damper coeffi-

cients so that the door opens enough with the applied force for a person to pass through, and closes in a reasonable amount of time.

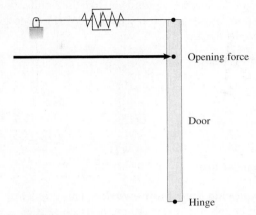

Figure E10.2

10.3 For Fun and Further Challenge Using Figure E10.3 as a guide, modify your simulation from Exercise 8.2 to add a damper in parallel with the spring and have the spring always active. Can you do a better job of reducing the maximum force on the train and controlling the overall motion than with the spring alone?

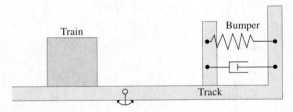

Figure E10.3

10.4 Design and Analysis Build what is called a "pitch-heave" or "bicycle" model of a vehicle, as shown in Figure E10.4. The vehicle body is a 500-kg polygon, the suspension arms are 20-kg rectangles pinned to the vehicle body, and the wheels are

10-kg circles. The front wheel is pinned to the front suspension arm and the rear wheel has a 100-N/m torque motor connecting it to the suspension arm. The spring and damper constants are 20,000 N/m and 1400 N-s/m, respectively. Put the vehicle on a custom ground and run the simulation. Make any adjustments necessary in terms of design and rerun the simulation until you like the results. Report on what you learned about Working Model and the dynamics of the vehicle.

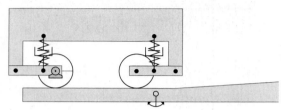

Figure E10.4

Exploring Your Homework   This problem can be solved with energy methods a bit easier than with Newton's laws.

10.5   The 20-kg cylinder shown in Figure E10.5 is released at the position shown and falls onto the linear spring ($k = 3000$ N/m). Use conservation of energy to determine how far down the cylinder

moves after contacting the spring. In Working Model, you will have to connect the top end of the spring to a small mass of about 0.5 kg, and the bottom end to the background. If each end were connected to the background, the cylinder would not contact the spring, but would simply fall past it.

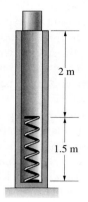

Figure E10.5

Exercise 10.5 is adapted from Bedford and Fowler's Engineering Mechanics: Dynamics *(Addison-Wesley, 1995), page 171.*

Basic Kinematic Linkages

Objectives

In this tutorial you will learn to:
• Reshape polygons
• Use the Smart Editor
• Use the Slot tools
• Rotate objects

Design principles related to this tutorial

• Kinematic analysis of the four-bar linkage
• Coupler points and curves
• Kinematic analysis of the slider-crank linkage

Introduction

A vast number of machines are created from only three basic kinematic linkages: the *four-bar*, the *slider-crank*, and the *double-slider crank*. Steering mechanisms in vehicles, windshield wiper systems, and lawn sprinklers are only a few examples of mechanisms that can be created from four-bar linkages. Slider-crank linkages are used in the design of internal combustion engines and air compressors.

Problem Statement

Figure 11.1 shows schematics of the four-bar and the slider-crank linkages, and the dimensions to be used for this tutorial. You will create Working Model simulations for both types using the dimensions shown in the figure as follows:

Create the four-bar linkage using an input angular velocity of ω_{AB} = 2 rps counterclockwise. Show the path of point C as this mechanism moves, and output a graph of the coordinates of this path. Also, show a graph of the ratio of output angular velocity to input angular velocity (ω_{DE}/ω_{AB}).

Create a slider-crank linkage using the same input angular velocity, ω_{AB}. Show graphs of the position, velocity, and acceleration of slider D as functions of time. Rotate the path of the slider 45° clockwise, and show the resulting position, velocity, and acceleration of the slider.

All sliders are 0.4 m × 0.6 m.

Figure 11.1

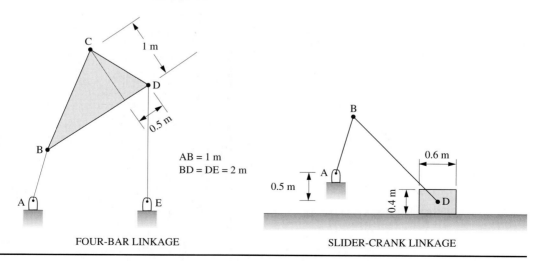

AB = 1 m
BD = DE = 2 m

FOUR-BAR LINKAGE   SLIDER-CRANK LINKAGE

Creating the Components

- **Open** a new, untitled Working Model window, with the default workspace you have used in each tutorial. The unit system and the accuracy mode should be the Working Model defaults, SI(degrees) and Accurate.

You will start by creating the four-bar linkage.

- Create two **rectangles** with the following dimensions:
  1. height = **0.1** m; width = **1** m
  2. height = **0.1** m; width = **2** m

These rectangles represent bars AB and DE.

- Use the **Polygon** tool to create an isosceles triangle with base = **2** m and height = **1** m. Its lower-left vertex should be located at (**-1,-2**).

- Move the lower-left-hand corner of the smaller rectangle to the **origin**.

- Place a **motor** near the origin end of the smaller rectangle.

  Note: You may need to turn off Grid Snap to place the motor on the bar. Remember to turn it back on when you have finished.

Your workspace should look like Figure 11.2.

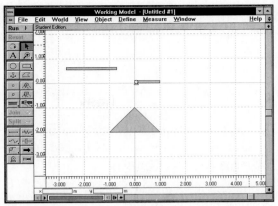

Figure II.2

- Open the **Properties** window for the motor.

The Properties window indicates the identities of the two points associated with the motor. In this case it shows that the base point is Point[5] and the motor point is Point[6], as shown in Figure 11.3.

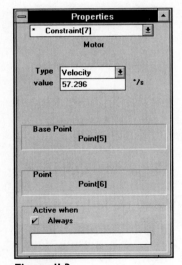

Figure II.3

- Change the value of the motor's **velocity** to **720°**/s (2 rps).

- Look at the Properties window for the **base point** of the motor.
- Change the **x, y** coordinates of the base point to (**0,0**).

Your bar and motor will move slightly in the workspace.

- Look at the Properties window for the **point** of the motor, which is located on the bar mass.
- Change the relative **x, y** coordinates of the motor **point** to (**-0.5,0**).
- Close the **Properties** window.

Your workspace should now look like Figure 11.4.

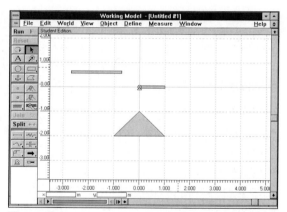

Figure 11.4

Reshaping Polygons

You can change the shape of a polygon after it has been created using the Reshape option from the Edit menu. You will use it to relocate the triangle's top vertex to create the shape shown in Figure 11.1.

- Select the triangle.

- Select **Edit** menu, **Reshape**.

Your cursor symbol changes so that it looks like a small box with crosshairs through it.

- Click and drag on the top vertex to move it **0.5** m to the right.

Note: Your coordinates display will show $\varnothing x = 0.5$, $\varnothing y = 0.0$ when you have dragged the vertex to the correct position.

- Select **Edit** menu, **Reshape** to turn off reshaping.
- Open the **Geometry** window for the triangle.

Examine the vertices at the bottom of the Geometry window shown in Figure 11.5. Make sure that your global coordinates are located as indicated in Figure 11.5. (The coordinates in your window may be lined up with different vertex numbers, depending on which point you selected first, second, and third.) You can change the coordinates here numerically, if you did not locate them precisely. Similarly, you can add vertices to (or delete them from) your polygon.

To add a vertex using the Geometry window, select the polygon, select a vertex that will be next to the new vertex, and click the Insert button. A duplicate vertex is created in the vertex list. The new vertex will move when you edit the coordinates of the duplicate vertex to create a geometrically distinct point. Similarly, you can delete a vertex in the Geometry window by selecting the polygon, selecting the vertex to be deleted, and clicking the Delete button in the Geometry window.

You can also add a vertex using the Reshape option. After selecting Edit menu, Reshape, click on the desired side of the polygon (not at an existing vertex), and drag the new vertex to the desired location. To delete a vertex using the Reshape option, select Edit menu, Reshape, click on the desired vertex, and select Edit menu, Clear. Remember to turn off reshaping each time you are finished editing a polygon.

Experiment with adding and deleting vertices on your own if you like.

> Note: A polygon must have at least three vertices; therefore you must add a vertex to your triangle before experimenting with deleting vertices now.

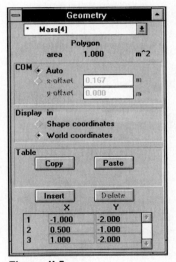

Figure II.5

- Use the **Point element** tool to place points near the two ends of the long bar, at the non-motor end of the short bar, and near the corners of the triangle.
- Select all of the points located on the bars.
- Use the **Properties** window to locate the points on the bars precisely at the center endpoints. These points will all need to have

the **relative** coordinate **y = 0.0**; you can change them all at once in the Properties window.

- Change the x-position components one at a time so that the points are attached to the very ends of the bars.

Tip Remember that to place points at the exact end of a mass, you must place them at the relative coordinate that is one-half the width or height of your mass. To move the point to the left or bottom of the mass, include the negative of the value. For example, the second point on the short bar should be located at relative coordinate x = 0.5 m (because the width of the bar is 1 m); the point has already been located at the mass's center of height, y = 0.

You now want to locate the points on the triangle precisely at the triangle's vertices.

- Select the triangle.
- Select its **Geometry** window.
- Make sure that **Display in World Coordinates** is selected.

If Display in World Coordinates is not selected, the locations of the vertices listed at the bottom will be with respect to the FOR (frame of reference) marked at the center of the triangle.

- Note that the coordinates for **vertex 1** are **(-1.0,-2.0)**.
- Use the rulers in the workspace to help you identify which point on the triangle is closest to vertex 1.
- Select the point in the triangle closest to vertex 1.
- Select the **Properties** window.

- Change the **global** coordinates to be equal to the coordinates of vertex 1, (**-1.0,-2.0**).

Your workspace should look like Figure 11.6.

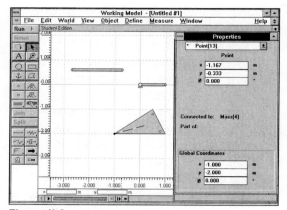

Figure 11.6

- Select the triangle and the **Geometry** window again.

- Note the global coordinates for vertices 2 and 3.

- Use the **Properties** window to move the point near vertex 2 to the global coordinates for vertex 2, and to move the point near vertex 3 to the global coordinates for vertex 3.

- Close the **Geometry** and **Properties** windows.

Your workspace should look like Figure 11.7.

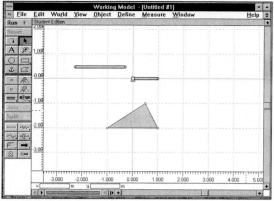

Figure 11.7

- **Join** the point on the short bar to the lower-left point on the triangle.

For a review of the Join tool, refer to Tutorial 7. The triangle moves to the bar because the motor is attaching the bar to the background. Your workspace should look like Figure 11.8.

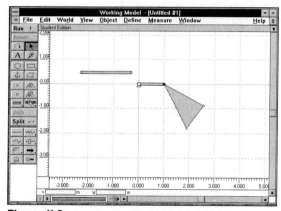

Figure 11.8

- **Join** the point on the triangle that is now the lowest point on your screen to either of the points on the long bar.

- Place a **point** on the background at (**2.0,0.0**).

Your workspace should look like Figure 11.9.

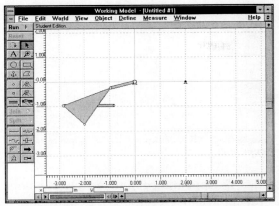

Figure 11.9

- **Join** this background point to the unattached point on the long bar.

Your workspace should look like Figure 11.10.

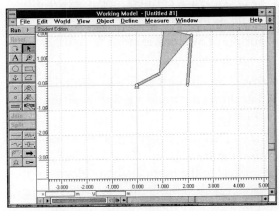

Figure 11.10

Using the Smart Editor

The Working Model Smart Editor allows you to manipulate objects and constraints while preserving the fundamental relationships that exist among them. "Manipulate" in this context has three possible meanings:

1. Dragging or rotating

2. Using the Join or Split tools

3. Typing values into the Properties window

You saw in Tutorial 10 how the Smart Editor prevents a mechanism from disintegrating when you move its components around. Instead, other components are moved or rotated (subject to their own constraints) until the desired move is accomplished. The Smart Editor is designed to follow the click-and-drag paradigm as much as possible. If a drag or rotation is inconsistent with a constraint, a compromise between the constraint and the drag is reached, but the constraint is always respected. When the Lock Points option (which you used in Tutorial 6) is in effect, it is impossible to drag the joints or endpoints of other constraints.

The rules that the Smart Editor uses in moving objects in the workspace are simple, consistent, and intuitive. The easiest way to understand the Smart Editor is to play with it. The following rules may help you to understand what you can and cannot do with the Smart Editor.

Rule #1: You cannot break a constraint during editing.

Rule #2: Endpoints of constraints (e.g., joints) cannot move on the objects they are attached to during editing.

Rule #3: If you simultaneously select a collection of objects, a drag or rotate operation treats them as a rigid unit, so that no alteration in their relative positions or relative rotations occurs.

Rule #4: Collisions are ignored during editing.

Rule #5: No joint rotates unless some constraint forces it to do so during editing.

If you try to perform an action that makes it impossible to satisfy all of the problem constraints, a warning box appears, informing you that Working Model cannot assemble the simulation, given the input requirements. Often, using the Split tool will enable you to edit the simulation as necessary. You will learn more about the Split tool in this tutorial.

- Click and drag to move the joined objects until your workspace looks like the four-bar linkage illustrated in Figure 11.1.

> **Tip** If you find it impossible to arrange your elements as they appear in Figure 11.1, select your joints and use the Split tool; it will enable you to move the pieces, as shown in Figure 11.11, to the approximate places that you want them to be. Then reselect the points of the joints and use Join to reconnect them, as shown in Figure 11.12. You can then click and drag on the short bar, as needed, to rotate it into approximately its position in Figure 11.1.

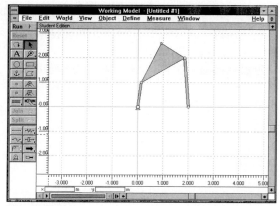

Figure 11.12

- Use the **Appearance** window to label points (or constraints) A through E as shown in Figure 11.13. (Use a selection rectangle as needed to select points without moving them. If you inadvertently move a point, select Edit Menu, Undo.)

- **Select all** objects and turn off **Track center of mass, Track connect,** and **Track outline** in the **Appearance** window.

- Turn on **Track connect** for the path of point C *only*.

- Create an output **meter** with **digital** display that shows the **x** and **y graph positions** of point C.

Now you will add a meter that shows the ratio of the angular velocity of the long bar (called the *rocker*) to the input angular velocity of the short bar (called the *crank*). You will start by adding a velocity meter for the rotation (angular velocity), then modify the angular velocity variable to represent ω_{DE}/ω_{AB}.

> **Note:** Working Model does not have a ω character available, so you will use a V to represent angular velocity in your label.

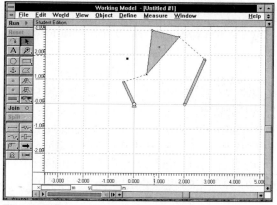

Figure 11.11

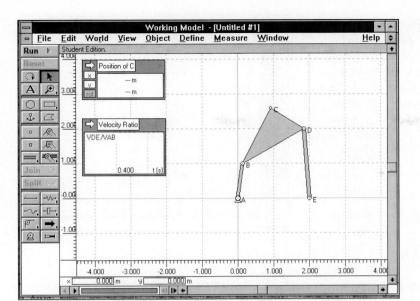

Figure 11.13

For a review of entering a formula in a meter's Properties window, refer to Tutorial 2.

- Create a **rotation graph velocity** output **meter** for bar **DE** in **graph** display.

- Modify the label and the formula for the velocity meter so that the graph represents ω_{DE}/ω_{AB} = mass [2].p.r/mass [1].p.r

- Label this meter **Velocity Ratio**, and change the color for the ω_{DE}/ω_{AB} curve to **red**.

Your workspace should look like Figure 11.13 when you are done.

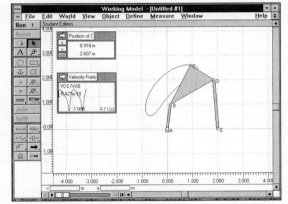

Figure 11.14

Running the Simulation

- **Run** the simulation through two complete rotations of the crank.

- Click **Reset**.

The curve drawn by connecting the track of point C is shown in Figure 11.14.

Link to Design

The body that connects the crank to the rocker is called the *coupler* (in the present case it is a triangle), and a selected point on the coupler (in this case, the top point on the triangle, point C) is called the *coupler point*. The curve drawn by the track of the coupler point, shown in Figure 11.14, is very often used as a design curve for mechanisms; it is called the *coupler curve*. For example, if you want a straight-line output motion, you would try to select dimensions for the mechanism so that a portion of the coupler curve would be a straight line. In this problem, a portion of the curve does look like a straight line. If desired, you could export the coupler curve data to a spreadsheet, or to another program, as you learned in Tutorial 9.

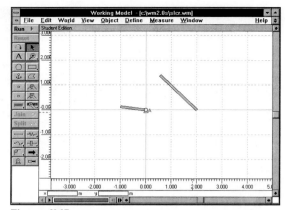

Figure 11.15

This completes the four-bar linkage portion of the simulation.

■ **Save** this simulation as **4BAR**

Creating the Slider-Crank Linkage

You will use this same simulation to develop the slider-crank linkage.

■ Use **Save As** to save the simulation again as **SLCR**

You will now change this four-bar simulation into a slider-crank simulation. Because the Working Model Smart Editor remembers all the previous joints formed in your simulation, you need to delete all the pins; you want to keep only the two bars and the motor from the four-bar linkage.

■ Select the two meters and the triangle.

■ **Clear** the selection.

The remaining pins, B, D, and E, are now attached to the background.

■ Select pins **B, D,** and **E,** and point **C.**

■ **Clear** the selection.

Your workspace should look similar to Figure 11.15. If you have made a mistake, select Edit menu, Undo, and try again.

■ Create a **rectangle 0.4 m** high × **0.6 m** wide.

■ Use the **Point** tool and the **Properties** window to place points at the three free ends of the bars where they were before.

■ Place a **point** at the center of the new rectangle.

These new points will be able to create new joints; they are not hampered by old Working Model constraints. Be sure the points are located on the bars (they should look like circles) and not the background (they would appear to have a triangular base). Your workspace should look like Figure 11.16.

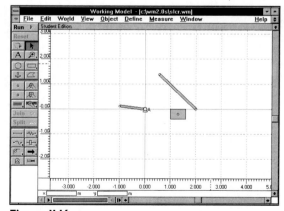

Figure 11.16

- **Join** the point on the short bar to a point on the long bar.

Using the Slot Tools

You create slots using the Slot icon on the tool bar in the left-hand column above the Join button. If you click and hold down on the Slot icon, you will see four different Slot tools: horizontal, vertical, curved, and closed curved. In this tutorial, you will use the horizontal slot. Curved slots will be discussed in Tutorial 12.

- Click on the **horizontal Slot** icon on the tool bar (in the left-hand column above the Join button).
- To create a horizontal slot, click at **x = 1.0** m, **y = -0.5** m.

Your workspace should look like Figure 11.17.

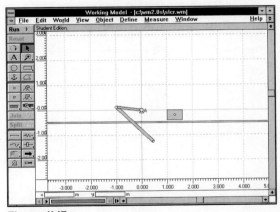

Figure 11.17

- With the slot already selected, hold down the **Shift** key and select the point on the slider rectangle.
- **Join** them.

Your workspace should look something like Figure 11.18.

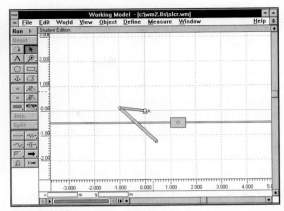

Figure 11.18

To connect another object on top of an already existing joint you need another free point that it can be joined to. You want to connect the long bar to the slider rectangle as well, so you will place another point at the center of the slider rectangle.

- Place another **point** precisely at the center of the slider rectangle.
- **Join** this new point with the point at the free end of the long bar.
- Click and drag on the crank until the workspace looks like the slider-crank linkage in Figure 11.1, using Split and Join if necessary.
- Label the point connecting the two bars **B**, and the slider rectangle **D**.
- Open a **P-V-A meter** for the **x** component of the slider rectangle in **graph** display.
- Name and position the meter appropriately, turn on the grid for the meter, and size the meter so that it will be large enough to look at the movements of the slider in detail.

Your workspace should look like Figure 11.19. The time scale for the meter will be smaller than that shown in the figure until you run the simulation, unless you adjust the maximum value for time, *x*, in the meter's Properties window.

Running the Simulation

- Turn **Tracking** off using the **World** menu.
- **Run** the simulation through approximately four cycles of the crank rotation.
- Click **Reset**.
- Save **SLCR**

Link to Design Note that the position or displacement graph in Figure 11.20 looks something like a sine wave; it is, in fact, close to a sine wave (simple harmonic motion), and gets closer as the length of the connecting rod (the longer bar) is increased. The velocity and acceleration curves are the first and second derivatives of the displacement curves.

The curves of the graph may be drawn to different scales. You can use the Properties window to set the y values for each curve equal to one another. This will give you a more accurate, easy-to-read graph.

To examine the actual numbers in detail, you can export the data to a spreadsheet, as you did in Tutorial 9.

Figure 11.19

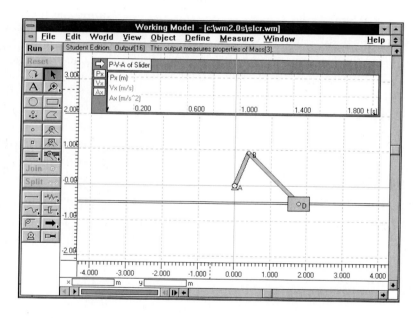

springs

15.628

7.814

25.25

35.276

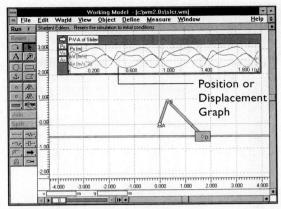

Figure II.20

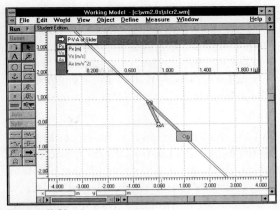

Figure II.2I

You will now use this simulation to experiment with rotating objects.

- Use **Save As** to save your simulation as **SLCR2**

Rotating Objects

There are two ways in which you can rotate objects: you can use the Properties window or the Rotate tool. You will now rotate the horizontal slot using the Properties window.

- Open the **Properties** window for the slot.
- Change the angle to **−45°**.
- Close the **Properties** window.

Your workspace should look similar to Figure 11.21.

- **Run** the simulation again.
- Click **Reset**.

Although there is now a y component to the position, velocity, and acceleration of the slider, the curves in Figure 11.22 represent only the x component, which is considerably less than when the slot was horizontal. To represent the total displacement, experiment with changing the equation in the position output meter or simply open another output meter for the total position of the slider.

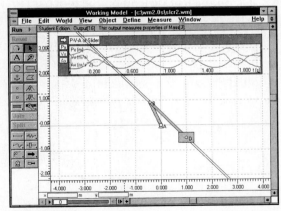

Figure II.22

You can also execute rotations using the Rotate icon. The Rotate tool allows you to rotate selected objects about a given point. The point chosen to be "rotated about" is the base point. The base point can be any point in the workspace, e.g., a pin joint or a center of mass.

You use the Rotate tool first to select objects for rotation, then use it again for the actual rotation operation. After you select the Rotate tool, select the object(s) you wish to rotate. When you do, a line snaps from the pointer to the closest possible rotation point in the workspace. This is the base point about which the object will rotate. Clicking close to the point on the object about which you wish to rotate it is most efficient. You then drag the object to rotate it.

Rotating is another function of the Smart Editor. As in the case of moving objects, rotating an object that is connected to other objects by constraints may cause these other objects to rotate, or move in some way, as well. You will now rotate the slider using the Rotate tool.

- Click on the **Rotate** icon on the tool bar (to the left of Arrow).

- Move the cursor over objects in the workspace.

A dotted line appears between the cursor arrow and the nearest base point.

- Click and drag on slider **D** to rotate it.

Your workspace should look similar to Figure 11.23.

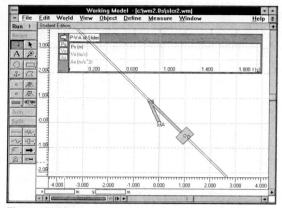

Figure 11.23

In some cases you may want to rotate the selection about a point that is not closest to the pointer. It is possible to fix a base point for rotation, rather than using the nearest default base point. You can do this by selecting a base point with the Rotate tool and holding down the **W** Ctrl key or **M** Command key while moving the mouse to whatever object you want to rotate about the chosen base point. The base point must still be a joint or a mass center; you cannot rotate an object about a random point on an object or the background.

By selecting a group of objects, you can rotate them all together with the Rotate tool.

- Click on the **Arrow** icon to deselect Rotate.

- **Save SLCR2** and **exit** Working Model.

11.1 The Basics Make a horizontal slot and rotate it 30°. Make a bucket similar to the one in Figure E11.1 with the Polygon tool, add a point to its top, and join it to the slot. This models the pre-launch escape system from the Gemini spacecraft. Create meters that show the angular position of the bucket and its vertical velocity. Run the simulation and monitor the motion. Select Reshape from the Edit menu to change the shape of the bucket, keeping its area (shown in the Geometry window) and its mass approximately constant. Does the shape affect the translational and rotational motion?

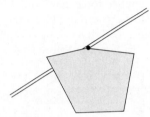

Figure E 11.1

11.2 Applying the Basics In the Robotics and Automation Laboratory, you are building a Cartesian robot, one that moves along the x and y axes independently. To model this in Working Model, create two squares labeled A and B to represent the moving platforms and place a square point at each center. (When square points connect objects to other objects or slots, they allow no rotation; you will learn more about this in Tutorial 12.) Place a horizontal slot on the x axis of the background and join it to the square point on platform B. Place a vertical slot directly on platform B, as shown in Figure E11.2. (Slots that are not created over the background attach themselves to the

object they are created over at the point you select when creating them.) Join the vertical slot to the square point on platform A. Attach a linear velocity actuator between platform B and the ground with velocity cos(3*t). Attach another velocity linear actuator between platforms A and B, as shown in Figure E11.2. Give this actuator a velocity of cos(5*t). Track only the center of mass of A. (Set the pattern of platform A to None in the Appearance window. This way, platform A will not obscure the path of its center of mass.) For what types of applications might a Cartesian robot be applicable?

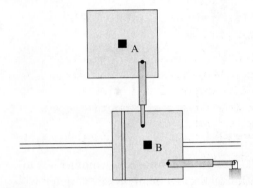

Figure E 11.2

11.3 For Fun and Further Challenge Modify your simulation SLCR from this tutorial to create a double slider-crank linkage, as shown in Figure E11.3. Create a new vertical slider, slider B (with the same dimensions as slider D), and a vertical slot on the y axis. Join slider B, the vertical slot, and the bar. Let this simulation be driven by gravity. Show a graph displaying the ratio of the horizontal position of slider D to the vertical position of slider B (x_D/y_B). (At certain times the denominator of this ratio will be zero. To avoid this, you will insert a conditional expression:

if(Mass[5].$p.y$ = 0,0,Mass[3].$p.x$/Mass[5].$p.y$)

which will set the ratio equal to zero if the denominator is zero.) Run the simulation. The vertical slider returns each time to its initial position. What does this tell you about the friction coefficients of the slots? Place a new reference frame on the vertical slider. How does the motion change?

Rotate the horizontal slot 30° counterclockwise. How does the motion change?

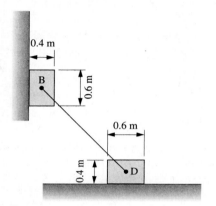

Figure E II.3

II.4. Design and Analysis   Another way to help model the motion of the suspension is with slots. Build a 520-kg vehicle body with the Polygon tool, as shown in Figure E11.4, and edit it until you like its looks. Select the Slot tool and place vertical slots at the positions of the front and rear axles. If you click on an object while a slot tool is selected, the slot automatically attaches to the object rather than to the background. The two slots are now attached to the vehicle instead of to the background. Create two 35-kg tires with points in the centers; join the center points to the slots so the wheels can only move vertically with respect to the vehicle. Add 20,000-N/m springs and 1400-N-s/m dampers between the wheel centers and the vehicle body. Build an uneven ground with the Polygon tool and place the vehicle model on it. (You may need to run the simulation under the influence of gravity only to let the car settle on the road, then use Start Here to set a new reset position.) Run the simulation with an applied force to the car of between 200 and 2000 N. Examine the effect of the applied force position (high or low on the vehicle body) and the applied force magnitude. Note: You may want to use the Zoom out tool after creating the car to create a road that is long enough to get up to speed on, as shown in Figure E11.5.

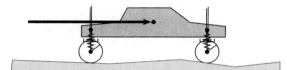

Figure E II.4

Figure E II.5

Exploring Your Homework This problem includes slots and pulleys. It can be solved easily by hand with energy methods. Compare a hand solution to that of Working Model.

II.5 The system is released from rest in the position shown in Figure E11.6. The weights are $W_A = 40$ lb and $W_B = 300$ lb. Neglect friction. What is the magnitude of the velocity of A when it has risen to 4 ft?

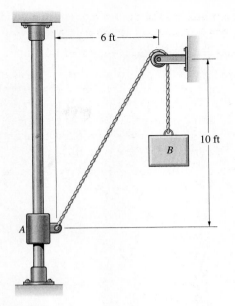

Figure E II.6

Exercise 11.5 is adapted from Bedford and Fowler's Engineering Mechanics: Dynamics (Addison-Wesley, 1995), page 159.

Cam-Follower
System

Objectives

In this tutorial you will learn to:
- Create curved slots
- Copy geometry from other applications
- Create keyed slot joints
- Use the Send To Back option
- Reshape curved slots

Design principles related to this tutorial

- Cam-follower systems
- Base circle
- Cam profile
- Follower displacement, velocity, and acceleration diagrams
- Parabolic follower motion
- Simple harmonic follower motion
- Cycloidal follower motion

Cams are mechanisms whose shapes create a program that produces a certain type of output motion; the cam's output device is called the *follower*. Cams are used in a wide variety of machines, e.g., timing systems in automobile engines, valves, and machine tools. In this tutorial you will analyze the output of a given cam design, using Working Model. You will also learn how to copy external data files and use the data to shape curved slots.

Problem Statement

The radial disk cam shown in Figure 12.1 has been designed to output parabolic motion to a roller follower. The base circle of the cam (the basis for the design of the cam, and the smallest circle which will fit on the cam) is 4 in., and it is rotating at 2 rps (180°/s). The coordinates of the cam profile (i.e., the shape of the cam) are given in Table 12.1 for every 10 degrees of rotation (for an actual cam, many more data points would be used in order for the cam surface to be smooth and continuous). Use this data to create a Working Model simulation of the cam follower system. Show follower displacement, velocity, and acceleration diagrams for the system.

Figure 12.1

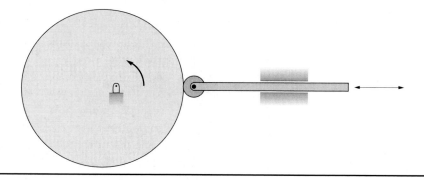

The data from Table 12.1 has been included in a spreadsheet named W CAM1.XLS or M CAM1, in the datafiles provided with your Working Model software. W CAM1.XLS or M CAM1 is a Microsoft Excel file. This data describes the shape, in polar coordinates, of the cam. If you are not running Microsoft Excel, you can create a table in almost any spreadsheet or word processing software and bring it into Working Model by copying it to the Clipboard, and then pasting it into Working Model. Your spreadsheet should be arranged (without the table title and column headings) as shown in Figure 12.2. The display of the spreadsheet in Figure 12.2 is set to only 2 decimal places, but you should enter the data with the accuracy shown in Table 12.1.

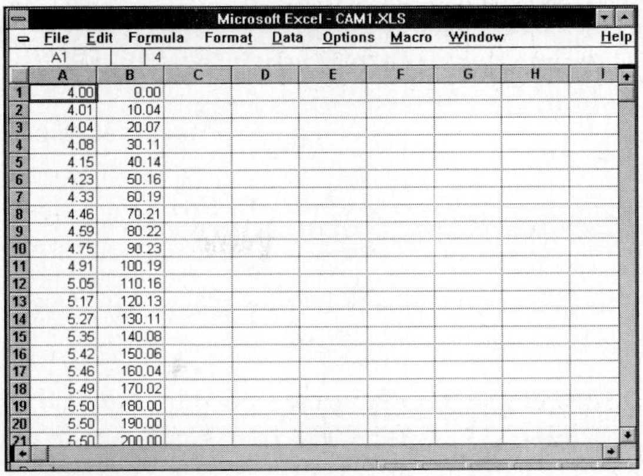

	A	B
1	4.00	0.00
2	4.01	10.04
3	4.04	20.07
4	4.08	30.11
5	4.15	40.14
6	4.23	50.16
7	4.33	60.19
8	4.46	70.21
9	4.59	80.22
10	4.75	90.23
11	4.91	100.19
12	5.05	110.16
13	5.17	120.13
14	5.27	130.11
15	5.35	140.08
16	5.42	150.06
17	5.46	160.04
18	5.49	170.02
19	5.50	180.00
20	5.50	190.00
21	5.50	200.00

Figure 12.2

Starting Working Model

- **Open** a new, untitled Working Model window, with the default workspace you have used in each tutorial. The accuracy mode should be the Working Model default, Accurate.

Table 12.1 Parabolic Cam Profile Coordinates

r	θ	r	θ	r	θ	r	θ
4.00000	0.00	4.90883	100.19	5.50000	200.00	4.48199	299.75
4.00929	10.04	5.04734	110.16	5.50000	210.00	4.33482	309.78
4.03717	20.07	5.16740	120.13	5.48670	219.97	4.21435	319.81
4.08363	30.11	5.26901	130.11	5.44682	229.94	4.12060	329.85
4.14866	40.14	5.35216	140.08	5.38035	239.91	4.05361	339.90
4.23224	50.16	5.41684	150.06	5.28732	249.88	4.01340	349.95
4.33437	60.19	5.46304	160.04	5.16772	259.84	4.00000	360.00
4.45504	70.21	5.49076	170.02	5.02160	269.80		
4.59422	80.22	5.50000	180.00	4.84897	279.75		
4.75192	90.23	5.50000	190.00	4.65582	289.73		

- Use the **Numbers and Units** dialog box to change the units to **English (slugs)**, and the numbers to **Fixed Point**, 5 digits.

Creating the Components

You will create a 4-ft radius circle to serve as the base circle of the cam.

- Create a **circle** of radius **4** ft, centered at the origin.

  Tip You can create the circle and use the coordinate display to size and place it. Click on the Circle icon, then click and hold at (-4,-4). Drag to ⌀x = 8, ⌀y = 8.

Your workspace should look like Figure 12.3.

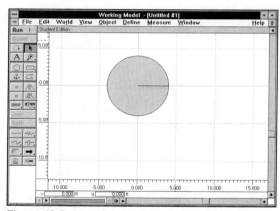

Figure 12.3

- Place a **motor** at the mass center of the circle.

- Use the **Properties** window to define the **velocity** of the motor as **180** °/sec.

The Properties window should look like Figure 12.4.

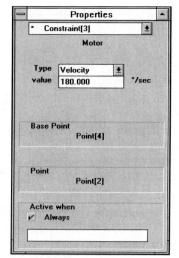

Figure 12.4

Creating Curved Slots

There are two types of *curved slots*; their icons pop up when you click and hold the horizontal Slot tool that you used in Tutorial 11. You use the curved slot element to create a curved slot from a series of smoothly interpolated control points. You define each control point with a single click, and use a double-click to identify the last point. You use the closed curved slot element to create a slot consisting of a closed curve. You define each control point with a single click, and use a double-click to identify the last point and automatically close the curve.

- Click and hold down on the **Slot** icon on the tool bar.

- Select the **closed Curved Slot element** (it looks like a doughnut).

- Click at three points on the circle: one point above, one to the right, and one below the motor.

- Double-click to the left of the motor to close the slot.

Your workspace should look something like Figure 12.5. It doesn't matter what the slot looks like; you will copy the data from the spreadsheet CAM1.XLS for the exact cam shape.

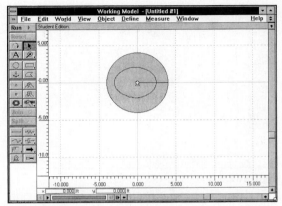

Figure 12.5

Copying Geometry from Other Applications

In Tutorial 9 you learned how to copy data from a Working Model simulation to another application. Bringing data from another application into Working Model is quite similar. In this tutorial you will see how to copy the closed curve slot geometry from a spreadsheet. You can use the same process for copying open curved slots and polygon geometries from external files. Slots and polygons are

transferred via the Clipboard as coordinates of the vertices. Specifically, the data is simple text consisting of a list of coordinate pairs, (x_1, y_1), (x_2, y_2) ... (x_n, y_n), delimited by tabs. Each number pair is on a separate line.

- **Save** your simulation as **CAM1.WM**.

Tip It is always a good idea to save your file before switching between applications.

W Open Microsoft Excel and the spreadsheet CAM1.XLS.

M Open Microsoft Excel and the spreadsheet CAM1.

- Make sure that the first column of your spreadsheet is set to 5 decimal places.

- Select the data in the first two columns of the spreadsheet.

- Use the **Copy** command on the selected text (this copies it to the Clipboard).

- Close the spreadsheet.

Note: You may be prompted by a dialog box asking: Save large clipboard from **W** CAM1.XLS or **M** CAM1? You should respond Yes.

- Return to Working Model **CAM1.WM**.
- Select the curved slot.
- Open the **Geometry** window.
- Click on **Paste** in the **Table** field of the Geometry window.

The vertices listed at the bottom of the Geometry window now show the data from the spreadsheet, and the cam shape in the workspace is drawn as required. Your workspace should look like Figure 12.6.

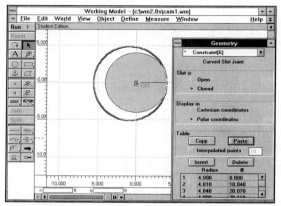

Figure 12.6

- Close the **Geometry** window.
- Create a **rectangle 1** ft high by **5** ft long to the right of your cam.
- Center the rectangle on the x axis.
- Place a **point** at **y = 0** on the rectangle near its left end.

Your workspace should look like Figure 12.7.

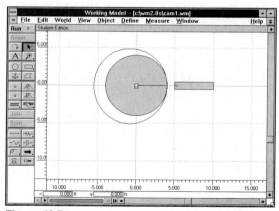

Figure 12.7

- Select the new point and the slot and **join** them.

Your workspace should look like Figure 12.8; if it does not, move the rectangle so that it is positioned as in the figure.

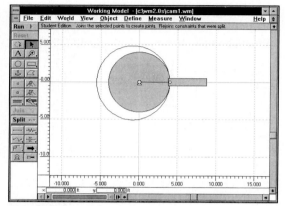

Figure 12.8

Creating Keyed Slot Joints

You can create slot joints by joining a pin and a slot together, as you did in the previous operation and in Tutorial 11. However, sometimes it is easier to create the entire slot joint in one operation by using the Slot joint tool. The Slot joint tool is located on the tool bar to the right of the Slot icon. There are five options under this icon: a horizontal pinned slot joint, a horizontal keyed slot joint, a vertical pinned slot joint, a vertical keyed slot joint, and a curved slot joint. A *pinned slot joint* is composed of a slot and a point. A *slot joint* aligns one point on one mass object with a slot on a second mass object or the background. A *keyed slot joint* is composed of a slot element and a square point element. It aligns a point on one mass object with a slot on a second mass object or the background, and prevents rotation between the two mass objects. When creating the next slot joint, you want to create a keyed slot joint; i.e., you do *not* want the rectangle (follower) to rotate relative to the slot that you will create.

- Click and hold down on the **Slot joint** icon on the tool bar (it looks like a round wrench on a horizontal bar).

- Select the **horizontal Keyed Slot joint** (it looks like a square wrench rather than a round one).

- Click at **y = 0** on the rectangle near its right-hand end.

Your workspace should look like Figure 12.9.

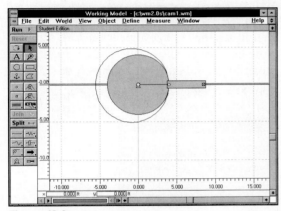

Figure 12.9

Using the Send To Back Option

Each object in the workspace can be moved in front of or behind other objects. You will move the follower to the back of the screen so that you can view the entire cam profile unobstructed in the workspace.

- Select the follower rectangle.

- Select **Object** menu, **Send To Back**.

The follower is now positioned behind the cam, but still in front of its horizontal slot.

- Create output meters displaying the **x position, x velocity,** and **x acceleration** of the follower in **graph** display, and a meter for the **time** in **digital** display.

- Use your scroll bars and the **Appearance** window to arrange the workspace and label the meters appropriately.

Your workspace should look like Figure 12.10.

Running the Simulation

- **Run** the simulation through two complete rotations of the cam.

- Select **Reset**.

Occasionally when you run a simulation, a Run dialog box appears on your screen. This box lets you know which frame Working Model is calculating. You can leave the box there while the simulation runs, or you can choose Hide to hide the box while you watch your Working Model simulation run. You can choose Stop to stop the simulation where it is, too.

When you are finished, the workspace should look like Figure 12.11.

Sometimes, in practice, the cam surface and follower might lose contact with each other because of inertia forces. This cannot happen in your simulation, since Working Model models the cam surface as a slot, not a surface. If the circle had been drawn larger than the cam profile, it would be clear that separation could not occur. A larger circle, however, would no longer represent the base circle for the cam.

Figure 12.10

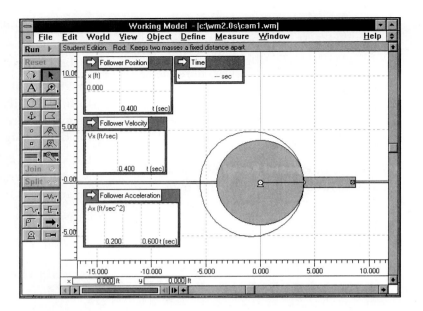

should see the change appear on your screen. Note how much clearer the displacement meter looks in Figure 12.13 than it does in Figure 12.11.

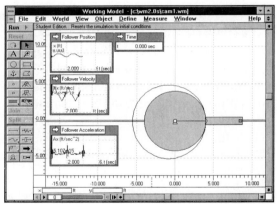

Figure 12.11

Obtaining Results

- Use the **Properties** window to change the scale for the displacement of y1 of the position meter to **5 < x < 8** so that the curve will fill the meter.

Your dialog box should look like Figure 12.12. When you have accepted these values, you

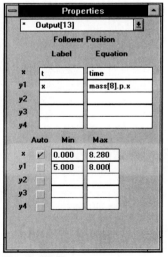

Figure 12.12

Figure 12.13

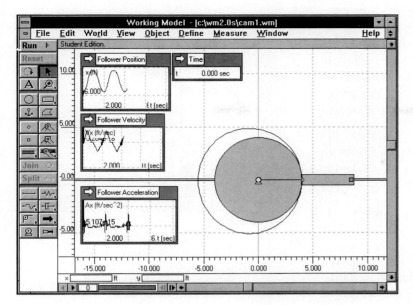

Figure 12.13

Link to Design

This cam shape was designed to yield a parabolic *follower motion*, i.e., a motion whose follower-displacement diagram is a parabola. You can tell that the top curve is a parabola by looking at its derivative, the follower-velocity curve; this curve consists of straight line segments. The acceleration curve should be constant (alternating between acceleration and deceleration). In fact, this type of cam output is referred to as motion with constant acceleration/deceleration. The reason that the acceleration curve looks so irregular is because, to save you time in entering the spreadsheet data, very few points were used to define the cam geometry. If many points had been used, both the velocity and acceleration curves would be much smoother.

- Use the **Tape Player** and switch between the **graph** display and the **digital** display of the acceleration meter to determine the maximum acceleration of the follower.

The maximum should occur at about $t = 2$ s, or frame number 100. The cam follower is very sensitive to changes in the cam's shape. Make a note of your maximum acceleration; you will now change the shape of the cam path and compare the resulting maximum acceleration.

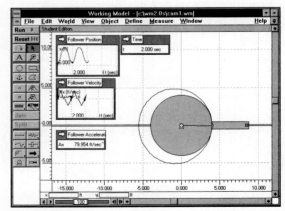

Figure 12.14

Reshaping curved slots

You can reshape curved slots using the Geometry window or the Reshape option on the Edit menu, in the same way that you reshaped polygons in Tutorial 11. You will change the shape of the cam to see what happens when the cam profile is more irregular.

- Return the acceleration meter to **graph** display.

- **Reset** the simulation.

- Select the curved slot (cam profile).

- Select **Edit** menu, **Reshape**.

Your workspace should look like Figure 12.15. Note that all the points that define the cam profile are displayed and the cursor is now displayed as a box with crosshairs in it.

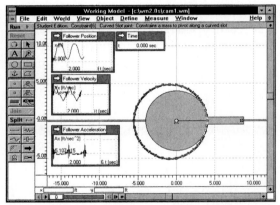

Figure 12.15

- Select the far left-hand point and move it to $\varnothing x = -1, \varnothing y = 0$.

Your workspace should look like Figure 12.16.

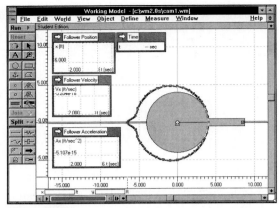

Figure 12.16

- Select **Edit** menu, **Reshape** to turn off the reshaping.

- **Run** the simulation for two complete revolutions again.

- Select **Reset**.

Your workspace should look like Figure 12.17.

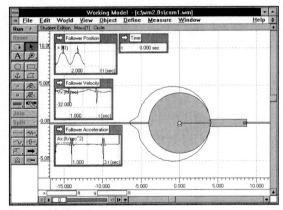

Figure 12.17

- Use the **Tape Player** and the **graph** and **digital** displays on the acceleration meter to observe the maximum absolute acceleration.

The maximum absolute acceleration is now extremely high and occurs at about frame 50, as shown in Figure 12.18. (Your maximum may be different.) You can see that the follower is extremely sensitive to changes in the cam shape; that is why many points on the profile need to be precisely defined to obtain a smoothly operating cam-follower system.

Tip You may want to use the Properties window to change the scale of the velocity and acceleration graphs so that you can see the complete curves in the meters.

■ **Save** this simulation as **CAM2.WM**

Link to Design

Tables 12.2 and 12.3 give data for cam profiles yielding simple harmonic motion and cycloidal motion. This data is in the datafiles provided with your Working Model software as Microsoft Excel files named **W** CAM2.XLS or **M** CAM2 and **W** CAM3.XLS or **M** CAM3, respectively. You can easily create cams for these motions by copying this data into your existing simulation. Compare the differences in acceleration behavior for these two types of cams with the parabolic cam, CAM1.

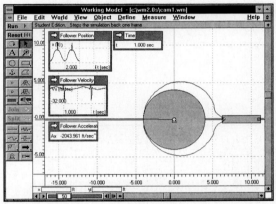

Figure 12.18

Table 12.2   Parabolic Cam Profile Coordinates

r	θ	r	θ	r	θ	r	θ
4.00000	0.00	5.17538	100.20	6.00000	200.00	4.69376	299.72
4.01528	10.06	5.34350	110.18	6.00000	210.00	4.50251	309.72
4.06065	20.12	5.50119	120.16	5.97823	219.96	4.33288	319.74
4.13468	30.16	5.64368	130.13	5.91388	229.92	4.19234	329.78
4.23506	40.20	5.76665	140.11	5.80973	239.88	4.08714	339.84
4.35869	50.22	5.86638	150.08	5.67033	249.85	4.02204	349.91
4.50176	60.24	5.93985	160.05	5.50171	259.81	4.00000	360.00
4.65991	70.24	5.98485	170.03	5.31121	269.78		
4.82833	80.23	6.00000	180.00	5.10711	279.75		
5.00190	90.22	6.00000	190.00	4.89826	289.73		

Table 12.3   Parabolic Cam Profile Coordinates

r	θ	r	θ	r	θ	r	θ
4.00000	0.00	5.22262	100.25	6.00000	200.00	4.61703	299.65
4.00225	10.01	5.42887	110.21	6.00000	210.00	4.39418	309.68
4.01768	20.05	5.61039	120.17	5.99614	219.99	4.21867	319.74
4.05797	30.11	5.75873	130.13	5.96997	229.96	4.09807	329.83
4.13175	40.17	5.86942	140.09	5.90312	239.91	4.03031	339.91
4.24356	50.23	5.94247	150.05	5.78426	249.86	4.00388	349.98
4.39324	60.27	5.98241	160.02	5.61099	259.80	4.00000	360.00
4.57601	70.29	5.99776	170.01	5.39018	269.74		
4.78314	80.29	6.00000	180.00	5.13685	279.69		
5.00304	90.28	6.00000	190.00	4.87157	289.66		

12.1 The Basics Starting from a new workspace, repeat the simulation of the tutorial for the harmonic cam profile given in Table 12.2. Based on the simulation's output, what are possible applications for this cam?

12.2 Applying the Basics Your job is to examine a scaled portion of a high-speed roller coaster, as shown in Figure E12.1. Use the English (slugs) unit system with gravity on. Make the track with the open curved Slot tool, the option on the Slot icon to the left of the closed Curved Slot tool you used in this tutorial. To make the track of the figure, select the open Curved Slot tool and click at (-10,5), (-5,0), (0,2.5), and (5,-2.5). Double-click at (10,-2.5). Make a 1-slug cart with a center point and join it to the slot. Set friction coefficients of all objects to zero. Move the cart to the first point of the track, create meters for gravitational potential and kinetic translational energy, and run the simulation. Can you explain what is happening to the system in terms of energy? What is true about the sum of the energies you are monitoring? What other physical effects might be important in actual design that are not represented in your simulation?

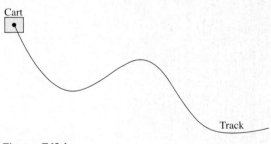

Figure E12.1

12.3 For Fun and Further Challenge
Create a figure-8 race track, as shown in Figure E12.2, with the closed Curved Slot tool, clicking at (-1,-1), (1,1), and (1,-3), and double-clicking at (-1,3). Use a circle to make a 2-kg "car" with a center point that is joined to the slot. Start the car at position (1,1) with an initial x velocity of 1 m/s and add a force of 100 N in magnitude such that the car completes the figure-8 in minimum time. Hint: Use a conditional statement to direct the angle of the force, based on the x and y velocity of the car. You will have to use the arc tangent function in Working Model, atan(•), where • is the function to evaluate. Use meters for velocity and time to help you.

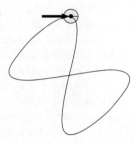

Figure E12.2

12.4 Design and Analysis The keyed slot joint will allow you to easily create a more realistic four-degree-of-freedom vehicle model. (This means that four motion coordinates are necessary to describe the motion.) The degrees of freedom are the vertical and rotational motion of the vehicle body and the independent vertical motion of the wheels.

Make a 450-kg vehicle body with the Polygon tool and add slot joints at the front and rear axle positions, as shown in Figure E12.3. You can use dimensions from your favorite car. Add square joints to the centers of the wheel squares and join them to the corresponding slots. Square joints will allow translation but not rotation of the wheels. Attach 15,000-N/m springs and 1200-N-s/m dampers between the vehicle body and wheel squares to represent a standard suspension. Create two small 5-kg rollers with point centers to represent ground-tire contact. Add additional slot joints on each of the wheel squares to guide the rollers. Join the roller centers with the new wheel slots. This will allow the rollers to translate vertically and rotate but not move horizontally with respect to the wheels. Connect a 170,000-N/m spring between each roller center and wheel square to model the spring characteristics of the tire.

Add a flat ground rectangle anchored to the background and under the vehicle, as shown in Figure E12.3. Run the simulation to allow the vehicle to settle on its suspension and stop bouncing (this may be slow, depending on the speed of your computer). Set a new starting point for the simulation with the Start Here option, set all initial velocities to 0, and give the vehicle body an initial vertical velocity of 1 m/s and a rotational velocity of 0.5 rad/s. Output the vehicle's velocities (vertical and rotational) and spring displacements and run the simulation. Can you decide how the responses of the model correspond to specifications for the performance of a real car? For instance, does the displacement of the tire spring represent an important behavior of suspensions?

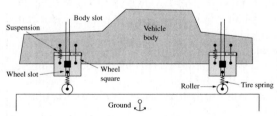

Figure E 12.3

Exploring Your Homework   This problem makes use of curved slots and springs. Energy methods can make an analytical solution easy. Compare a hand solution to that of Working Model.

12.5   The 5-lb collar in Figure E12.4 starts from rest at A and slides along the semicircular bar. The spring constant is $k = 100$ lb/ft, and the unstretched length of the spring is 1 ft. Use conservation of energy to determine the velocity of the collar at B.

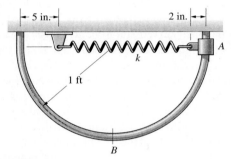

Figure E 12.4

Exercise 12.5 is adapted from Bedford and Fowler's Engineering Mechanics: Dynamics (Addison-Wesley, 1995), page 171.

Epicyclic Gear Train

Objectives

In this tutorial you will learn to:
- Create pin joints and rigid joints
- Use the Gear tool
- Assemble mechanisms that lie in parallel planes
- Use conditional statements
- Use chain drives and friction belts

Design principles related to this tutorial

- Compound gear trains
- Epicyclic or planetary gear trains
- Gear ratios
- Gear trains with more than one degree of freedom
- Chain drives and friction belts

Gear trains, chain drives, and friction belts are used in virtually every type of engineering application for transmitting rotational motion from one shaft to another. The ratio of output to input rotational speed is determined by the number of gears, their geometric arrangement, and their relative sizes. In this tutorial you will use Working Model gears to create a gear train; you will then examine how that gear train could be operated for several different applications.

Problem Statement

Figure 13.1 shows a schematic drawing of the top and side views of the pitch circles for a gear train composed of four spur gears. *Pitch circles* are imaginary circles that roll together as if the gears were friction wheels. The pitch circle radii are as follows: $r_1 = 0.5$ m, $r_2 = 0.8$ m, $r_3 = 0.2$ m, and $r_4 = 1.5$ m. Gear 4 is an internal gear. The arm, labeled object 5, is 1.3 m tall and 0.1 m wide. Gears 2 and 3 are rigidly keyed to the same shaft and rotate together. Gears 1 and 4 are attached to independent shafts. The gear train operates in a horizontal plane. Create a Working Model simulation of this gear train and examine the relative speed ratios of the gears for three cases:

1. The arm is fixed and gear 4 is driven at a constant speed of 30 rpm clockwise.

2. Gear 1 is fixed and gear 4 is driven at a constant speed of 30 rpm clockwise.

3. None of the objects is fixed, gear 4 is driven at a constant speed of 30 rpm clockwise, and arm 5 is driven at a constant speed of 10 rpm counterclockwise.

Figure 13.1

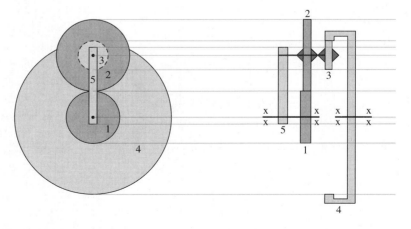

Setting Up the Workspace

- **Open** a new Working Model window with the workspace settings you have used in each tutorial.

- Be sure your simulation is using **Accurate** mode.

The gears operate in a horizontal plane; therefore there will be no acceleration due to gravity in this problem.

- Use the Gravity dialog box to set **Gravity** to **None.**

You will want to enter and view gear velocities in rad/s, so you will change the units now.

- Use the Numbers and Units dialog box to change the **Units Systems** to **SI (radians)**.

Creating the Components

You will now create and label the gears and arm. Since this is a fairly complicated mechanism, create the elements in the order in which they are numbered in Figure 13.1 and described in the Problem Statement, and assemble them exactly as instructed in this tutorial. The Working Model mass numbers will then be the same as the labels in Figure 13.1 and will correspond with the mass numbers given in the tutorial, so you will be able to check your work.

- Create four **circles** to the side of your workspace.

- Create one **rectangle** to the side of your workspace.

Your workspace should look like Figure 13.2.

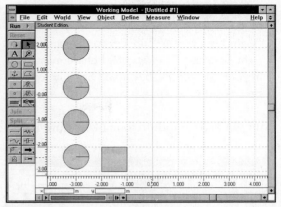

Figure 13.2

You will now begin to position and size the objects in your workspace. You *must* use the Properties window for the *exact placement of every object* in this tutorial. This is very important for smooth operation of the gears; you will not be able to run the simulation if you are not precise in the placement of all objects.

- Open the **Geometry, Properties,** and **Appearance** windows.

- Make the **radius** of Mass[1] equal to **0.5** m, and place its center at **(0,0)**.

- Make the **radius** of Mass[2] equal to **0.8** m, and place its center at **(0,1.3)**.

- Make the **radius** of Mass[3] equal to **0.2** m, and place its center at **(0,1.3)**.

- Label all five objects according to the Problem Statement, in the order in which they were created.

Tip You can include spaces in front of the label you type in the Appearance window to improve the location of the labels. In order to be able to see your labels once you have assembled this simulation, you should include 5 spaces before "1," 11 spaces before "2," 2 spaces before "3," 25 spaces before "4," and no spaces before "5." (Right now the "4" appears to be far to the right of circle 4 and may be behind mass 5 in your workspace, but it will be in its correct position when you resize circle 4.)

Your workspace should look like Figure 13.3.

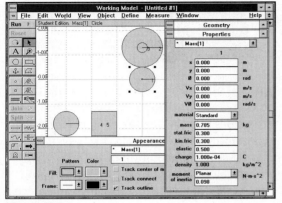

Figure 13.3

Creating Pin Joints and Rigid Joints

A *pin joint* acts as a hinge between two mass objects. A *rigid joint* prevents motion or rotation between two mass objects. Joints automatically connect the top two mass objects in the workspace. If only one mass object lies beneath a joint, then the joint connects the mass object to the background. You can create both types of joints either by clicking on the appropriate Joint icon in the tool bar and selecting the two overlapping objects to be joined in the workspace, or by placing two round or two square pins on two different objects and using the Join button as you have done before.

You will attach circle 1 to the background with a pin joint.

- Select the **Pin Joint** icon on the tool bar (to the right of the Point icon).

- Attach circle 1 to the background with a **pin joint** at the **origin**.

Your workspace should look like Figure 13.4.

Figure 13.4

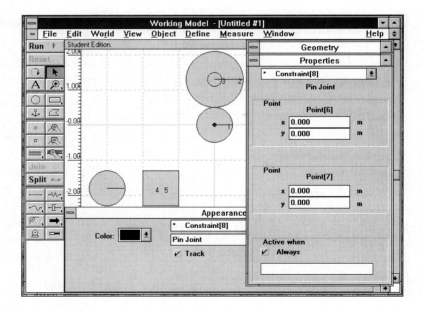

You will now join circles 2 and 3 with a rigid joint, which will force them to rotate together with the same angular velocity.

- Select the **Rigid Joint** icon on the tool bar (below the Pin Joint icon).

- Place a **rigid joint** at the exact center of circles 2 and 3.

Tip Leaving the Properties window open makes it easy to confirm the location of points and pins as they are created.

Defining rigid joint properties

The Properties window of a rigid joint (shown in Figure 13.5) has two radio buttons that specify whether the joint is *optimized* or *measurable*.

An optimized rigid joint neither introduces extra forces in the simulation nor affects simulation speed—the two rigidly connected objects behave as one. You cannot measure forces or torques on an optimized joint (the objects measure as 0.0).

To obtain correct force/torque readings, you must make the joint measurable. Because joined mass objects are treated individually, the simulation will take slightly longer to compute. Creating a force or torque meter for a rigid joint automatically makes it measurable (non-optimized).

Leave the rigid joint set to its default, optimized.

- Make the **radius** of circle 4 equal to **1.5** m, and place its center at **(0,0)**.

- Select **Object** menu, **Send To Back**, if necessary, so you can see all the objects in your workspace.

Note: You will be instructed to use the Object menu's Move To Front/Send To Back options frequently throughout this tutorial. If you need to use them to be able to work on your simulation more easily, use them freely at your own discretion.

Figure 13.5

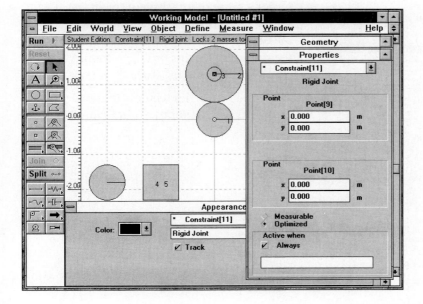

Your workspace should look like Figure 13.6.

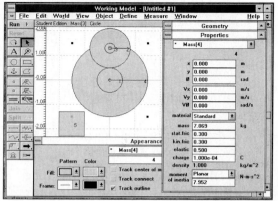

Figure 13.6

Adding the motor

You will now create a motor to drive gear 4 with the proper velocity. You will attach the motor away from circle 4's overlap with circle 1 so that the motor is attached to circle 4 and the background, not to circle 4 and circle 1.

- Attach a **motor** to circle 4 (attach it away from the overlap with circle 1).

Your workspace should look like Figure 13.7.

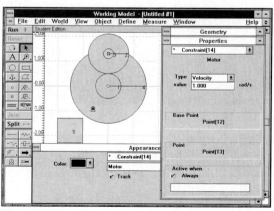

Figure 13.7

- Use the Properties window to move the motor's **base point** (Point[12] in Figure 13.7) to **(0,0)**.

This moves circle 4 away from the origin, as shown in Figure 13.8; the base point is where the motor attaches to the background. The motor and everything attached to it (circle 4) move to the new location.

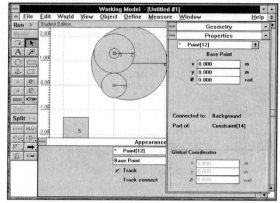

Figure 13.8

Now you will move circle 4 back to the origin, as shown in Figure 13.9, by changing the motor's point of attachment to circle 4. The motor point is where the motor attaches to the object, circle 4. Its location is relative to the circle and, in this problem, should be circle 4's center of mass, (0,0).

- Use the Properties window to move the motor's **point** (Point[13] in Figure 13.7) to **(0,0)**.

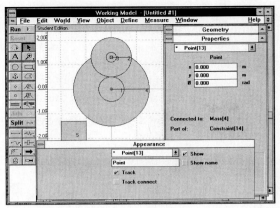

Figure 13.9

Now you will set the velocity of the motor. Working Model accepts "pi" as a value for the velocity of the motor, as shown in the Properties window in Figure 13.10. A clockwise rotation is negative in Working Model.

- Change the **value** for the velocity of the motor to **-pi** rad/s (-30 rpm).

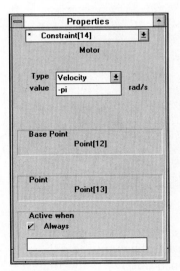

Figure 13.10

Tip Working Model is not case sensitive. Formulas may be typed with or without capitalization.

You will now start to assemble gears 3 and 4.

- Select circles 3 and 4.
- Select **Object** menu, **Move To Front**.

Using the Gear Tool

The *Gear* tool pops up from the Pulley System icon on the tool bar when you hold the Pulley System icon down. It provides a constraint on two mass objects so that their rotations are dependent on each other. When you define a pair of gears, the Working Model default is for a pair of external spur gears (i.e., gears that have teeth pointing radially outward). You have the option to define one of the gears as an internal gear (i.e., a gear that has teeth pointing radially inward). As explained in the Problem Statement, gear 4 is an internal gear.

The gear constraint in Working Model allows two rigid bodies to exert forces on each other at a single point of contact. The point of contact is located along a line passing through the centers of mass of the two bodies; its location depends on the gear ratio. For circular bodies, the gear ratio is computed as the ratio of the radii of the two bodies. For example, in this problem, the ratio of $r3/r4 = 0.2/1.5 = 0.133$; this will be the automatic gear ratio for gears 3 and 4 when they are created. Even if the two circles are not touching when the pair of gears is created, the automatic gear ratio will be the same, since it is based only on the ratio of the radii of the two circles. However, the theoretical point of contact would be somewhere in between the two circles, dividing the space between the two centers proportionally as the ratio of the radii.

The driving gear exerts a gear force on the driven gear in a direction perpendicular to the line of the centers (i.e., tangent to the pitch circles). Working Model computes the force necessary to maintain proportional rotation, angular velocity, and angular acceleration on both disks at the point of contact.

As mentioned above, Working Model gears need not be touching each other; once you have designated objects as gears, they behave as if they are touching.

Gears also need not be circular; they are automatically given a gear ratio of 1 if they are not circular. During the simulation, Working Model is only responsible for maintaining the rotations, angular velocities, and angular accelerations of the bodies in accordance with a given gear ratio. Working Model does not take into account the geometries of the objects. You can experiment with substituting a square for one of the circles when you have completed this tutorial.

All gear objects are automatically given the Do Not Collide designation. You cannot make the gear objects collide unless you remove the gear constraint between them.

Assembling mechanisms that lie in parallel planes

Although Working Model is a two-dimensional program, you can assemble systems that act in several planes (all of which are parallel to the Working Model workspace plane) by using the Split and Join tools and the Move To Front/Send To Back options. The gear train in

this tutorial has three parallel planes of motion: (1) the plane of gears 1 and 2, (2) the plane of gears 3 and 4, and (3) the plane of arm 5. The relative locations of the three planes (i.e., which one is on top, etc.) will not affect the relative motion of the gears and the arm.

- Click and hold down on the **Pulley System** icon on the tool bar.
- Select the **Gear** tool (to the right of the Pulley tool).
- Click on **circle 3** and drag to **circle 4**.
- Click on the **Arrow** tool to deselect the Gear tool (if necessary).

The order in which you create the gear is not important; you could click first on circle 4. The Gear tool always snaps the gear to the mass center of the objects. Your workspace should look like Figure 13.11.

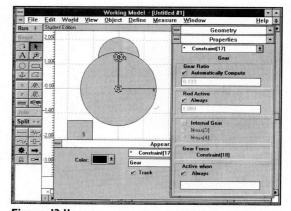

Figure 13.11

The top line of the Properties window in Figure 13.11 indicates the number for the gear set constraint: Constraint[17]. Below that is the Gear Ratio field; the Automatically Compute button is checked and the value 0.133 (the ratio of the radii of the two bodies) is shown grayed out in the box. If you turn off Automatically Compute, you can insert a different gear ratio.

The next field is the Rod Active field. The Rod Active field governs the existence of the rod connecting the gears, and behaves in the same way as the Active when field discussed in Tutorial 8. By default, each pair of gears has a rigid rod constraint between the two mass centers. The rod maintains a constant distance between the two objects. If you like, you can turn off the Rod Active Always option, and insert a formula describing conditions for the rod to be active.

The next field permits you to create an internal gear for one of the two gears. For this problem, you will make gear 4 internal.

- Check the **Internal Gear** option in the Properties window.

- Select **Mass[4]**, if it is not already selected.

Note: If one of the objects is designated as an internal gear, the gear ratio cannot be 1 unless the centers of mass of the two objects coincide. Otherwise, the point of contact would be located at infinity, and a warning dialog box would appear. A formula definition of the gear ratio cannot be evaluated until run-time, so if you include a formula, calculate the possible results to be sure that it does not return 1 for an internal gear ratio during the simulation.

Notice in Figure 13.12 that the gear symbol for Mass[4] now looks like an internal gear.

The next field in the gear's Properties window identifies the gear force as Constraint[18]. You can refer to this variable name to measure the gear force. In this problem, Constraint[18].f.y represents the force exerted by one gear on the

Figure 13.12

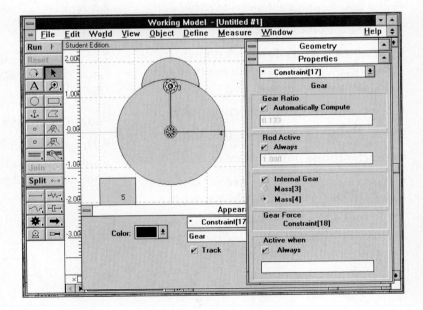

other (the x component is always zero). In addition, Constraint[17].f.x (Constraint[17] is the gear set) represents the force in the rod (the y component is always zero). You cannot open a Properties window for the gear force, Constraint[18], but you can use it in formulas. The last field in the Properties window is the Active when field and governs when the gear is active. It behaves in the same way as the Active when field discussed in Tutorial 8.

The rod created with a set of gears has no mass, so it cannot have torques applied to it or have an anchor or a motor placed on it. You will need to use rectangle 5 instead of the rod for these purposes to model the arm connected to gears 1 and 2.

- Change the dimensions of rectangle 5 to **height = 1.3** and **width = 0.1**.

- Close the **Geometry** and **Appearance** windows.

You will now use the point element to create round pins to join the objects together.

- Make the **Point** tool active.

- Attach **round points** to the endpoints of rectangle 5, relative coordinates **(0,-0.65)** and **(0,0.65)**.

- Select circle 1 in the Properties window.

- Select **Object** menu, **Move To Front**.

- Attach a **round point** to the center of circle 1.

- If necessary, click on the **Arrow** tool to deselect the Point.

- With the point on circle 1 already selected, hold down the **Shift** key and select the bottom point on rectangle 5.

- **Join** circle 1 and rectangle 5.

- If necessary, reselect rectangle 5 and use the **Move To Front** option so that you can see rectangle 5.

Rectangle 5 and circle 1 move together, as shown in Figure 13.13. The Smart Editor moves the objects to bring the two elements together as specified, while observing other existing constraints. To review the Smart Editor in more detail, refer to Tutorial 11.

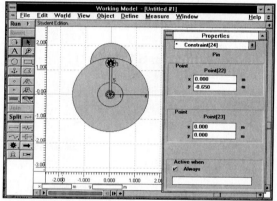

Figure 13.13

You will now connect rectangle 5 to circle 2.

- Drag the top end of rectangle 5 off to the side, so that its top point is not directly on top of circles 2 and 3.

Your workspace should look like Figure 13.14.

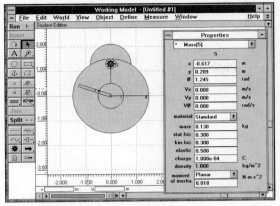

Figure 13.14

- Select circle 2 and rectangle 5.
- Select **Object** menu, **Move To Front**.
- Attach a temporary **anchor** to circle 2 (so it won't move when joined to rectangle 5).
- Attach a **round point** to the center of circle 2.
- With the point on circle 2 already selected, hold down the **Shift** key and select the free point on rectangle 5.
- **Join** circle 2 and rectangle 5.
- **Clear** the anchor.

Rectangle 5 and circle 2 have moved together and are now joined, as shown in Figure 13.15. (Your rectangle 5 may be behind circle 2.)

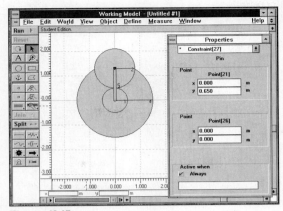

Figure 13.15

Now you will add the gear constraint for circles 1 and 2, then rearrange the objects for easier viewing.

- Select circles 1 and 2.
- Select **Object** menu, **Move To Front**.
- Select the **Gear** icon.
- Click on **circle 2** and drag to **circle 1**.
- Select circle 3.
- Select **Object** menu, **Move To Front**.
- Select rectangle 5.
- Select **Object** menu, **Move To Front**.

You have now assembled all the parts. Your workspace should look like Figure 13.16. Note that in their final form, gears 1 and 3 *seem* to be on the same plane, although this is not the case, because you created gear planes 1–2 and 3–4 as directed in the Problem Statement. As stated earlier, the relative locations of the three planes will not affect the relative motion of the circles (gears) and the rectangle (arm).

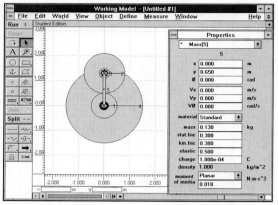

Figure 13.16

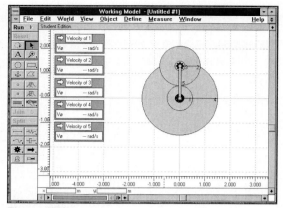

Figure 13.17

Although objects associated with the same gear constraint are automatically set to Do Not Collide, objects from different gear constraints may still collide. For example, gears 2 and 4 are not attached to the same gear constraint, so you must activate Do Not Collide for them.

- Select all objects in the workspace.
- Select **Object** menu, **Do Not Collide**.
- Select **View** menu, **Lock Points**.
- Create, label, and position output **meters** for the **rotation graph velocity** (angular velocity) of each object in **digital** display.
- Close the **Properties** window.

> Note: You will need to rename the meters to eliminate all the spaces you placed in the object labels. Use the **Move To Front/Send To Back** options as necessary to create the meters.

Your workspace should look like Figure 13.17.

Running the Simulation

You will now solve the first part of this problem, in which the arm is fixed. Anchors fix only the top item in the workspace; the objects beneath the arm will still be free to move.

- Place an **anchor** on arm 5.
- **Save** this simulation as **TRAIN1**
- **Run** the simulation through one complete rotation of gear 4.

All the gears rotate in place.

> Note: If you see the message: "Inconsistent constraints or physical instability has been detected. Some constraints may be ignored," it may be because you did not locate all of the points precisely as you developed the model. Table 13.1 shows the relevant values for all the masses, constraints, and points in the simulation. Check your Properties windows for all objects against Table 13.1 to see which ones need

Table 13.1   Relevant Values for Masses, Constraints, and Points

Mass object	Geometry	x position	y position	Global or relative coordinates	Angle	Base Point	Related Point	Related Point	Velocity value	Internal/external
Mass[1]	r = 0.500	0.000	0.000	Global	0					
Mass[2]	r = 0.800	0.000	1.300	Global	0					
Mass[3]	r = 0.200	0.000	1.300	Global	0					
Mass[4]	r = 1.500	0.000	0.000	Global	0					
Mass[5]	h = 1.300, w = 0.100	0.000	0.650	Global	0					
Constraint[8]							Point[6]	Point[7]		
Point[6]		0.000	0.000	Global	0					
Point[7]		0.000	0.000	Mass[1]	0					
Constraint[11]							Point[9]	Point[10]		
Point[9]		0.000	0.000	Mass[2]	0					
Point[10]		0.000	0.000	Mass[3]	0					
Constraint[14]						Point[12]	Point[13]		-pi	
Point[12]		0.000	0.000	Global	0					
Point[13]		0.000	0.000	Mass[4]	0					
Constraint[17]							Point[15]	Point[16]		External (Mass[3]) Internal (Mass[4])
Point[15]		0.000	0.000	Mass[3]	0					
Point[16]		0.000	0.000	Mass[4]	0					
Constraint[24]							Point[22]	Point[23]		
Point[22]		0.000	-0.650	Mass[5]	0					
Point[23]		0.000	-0.000	Mass[1]	0					
Constraint[27]							Point[21]	Point[26]		
Point[21]		0.000	0.650	Mass[5]	0					
Point[26]		0.000	0.000	Mass[2]	0					
Constraint[29]							Point[25]	Point[28]		both External
Point[25]		0.000	0.000	Mass[2]	0					
Point[28]		0.000	0.000	Mass[1]	0					

to be relocated. If you still see the message after having confirmed that your points, objects, and constraints are located precisely, Working Model is probably detecting an infinite acceleration at the first frame of your simulation. This will not affect your results. Choose "Continue" to allow your simulation to run.

Obtaining Results

Note that the angular velocity remains constant for all objects. The values for the angular velocities are shown in Table 13.2. Since the input angular velocity of gear 4 is constant, and the speed ratio is constant, all output angular velocities will also be constant.

A gear train whose arm is fixed and has two of its gears sharing a common shaft is known as a *compound gear train.*

Table 13.2

Input and Output Angular Velocities for Case 1	
	Case 1 (rad/s)
Gear 1	37.699
Gears 2 & 3	-23.562
Gear 4	-3.142 = -$\neq$
Arm 5	0.000

It would be interesting to see the ratio of the velocities of gear 1 to gear 4.

Using conditional statements

- Duplicate the meter measuring the velocity of gear 4.

- Move the meter to the bottom of the workplace.

- Use the Appearance window to label the meter **Ratio Gear 1: Gear 4**.

You now need to change the formula for the angular velocity (ω) to represent the ratio of the gear velocities, ω_1/ω_4. However, there will be times (at the start of the simulation) when the denominator of this ratio will be zero and the equation will result in an error. To avoid this you will insert a conditional expression which will set the ratio equal to zero if the denominator is zero.

The conditional expression has three parts: the condition to be tested ($\omega_4 = 0$), the value to be used if it is true (0), and the value to be used if it is false (ω_1/ω_4). All three parts are separated by commas and enclosed in parentheses. Using the Working Model variables for rotational velocity, this formula will look like this:

if(Mass[4].v.r = 0,0,Mass[1].v.r/Mass[4].v.r)

- Use the Properties window to change the formula for **V**ϕ to the conditional expression.

Your Properties window should look like Figure 13.18.

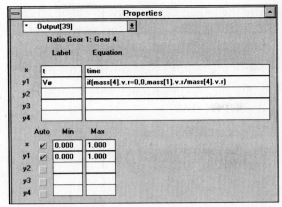

Figure 13.18

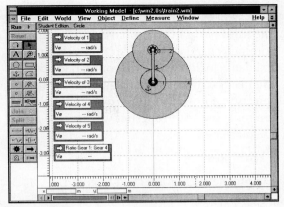

Figure 13.19

- **Run** the simulation through one complete rotation of gear 4.

- Select **Reset**.

The resulting ratio, -12, tells you the factor by which the input angular velocity is increased by the gear train. You can test Working Model's calculations in this simple ratio of output to input by using the values for the angular velocity in Table 13.2.

- **Save** this simulation as **TRAIN1**

Case 2

The second part of this problem requires that gear 1 be fixed.

- Clear the anchor attached to arm 5.

- Place an **anchor** on gear 1.

- Use **Save As** to save this simulation as **TRAIN2**

Your workspace should look like Figure 13.19.

- **Run** the simulation through one complete rotation of gear 4.

Gears 2 and 3 and arm 5 rotate with gear 4. Gear 1 does not rotate. Compare your results for Case 2 with the data shown in Table 13.3. The gear ratio for gears 1 and 4 is now zero since gear 1 is fixed.

Table 13.3

Input and Output Angular Velocities for Case 2	
	Case 2 (rad/s)
Gear 1	0.000
Gears 2 & 3	-4.712
Gear 4	$-3.142 = -\pi$
Arm 5	-2.899
Gear Ratio	0.000

A gear train with one gear fixed and a rotating arm is called a planetary or epicyclic drive with one degree of freedom.

Case 3

The third part of this problem requires that nothing is fixed and the arm rotates with an angular velocity of 10 rpm counterclockwise. You need to move arm 5 away from the other objects to connect a motor to it. Remember, you must use the Properties window for precise placement.

- Select **Reset**.
- Clear the anchor attached to gear 1.
- Select arm 5.
- Select **Split**.
- Drag arm 5 away from the other objects.
- Attach a **motor** to arm 5, using the Help Ribbon to position it on arm 5.

Tip If you have trouble getting the motor to attach itself to the arm, use the **Edit** menu, **Undo** to delete the motor, and try to position the motor on the arm again using the Help Ribbon as a guide.

- Use the Properties window to change the **value** of the velocity for the motor to **pi/3** rad/s (10 rpm counterclockwise).
- Move the motor's **base point** to the origin.
- Move the motor's **point** to the relative coordinates **(0,-0.65)**.

Your Properties windows for the motor, its base point, and its point, should look like Figure 13.20.

- Select arm 5.
- Select **Join**.
- Use **Save As** to save this simulation as **TRAIN3**
- **Run** the simulation through one complete rotation of **gear 4**.

Gears 2 and 3 rotate counterclockwise around gear 4. Gear 4 rotates clockwise. Compare your results for Case 3 with the data shown in Table 13.4.

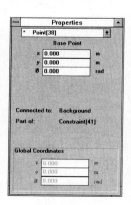

Figure 13.20

Table 13.4

Input and Output Angular Velocities for Case 3	
	Case 3 (rad/s)
Gear 1	51.313
Gears 2 & 3	-30.369
Gear 4	-3.142 = -≠
Arm 5	1.047 = ≠/3
Gear Ratio	-16.334

A gear train in which all elements can rotate must have multiple inputs; that is why you used two motors to specify two angular velocities for this problem: $\omega_1 = -≠$ and $\omega_4 = ≠/3$.

■ Select **Reset**.

Link to Design

The following equation is a formula that gives the relative angular velocities of gears 1 and 4 and arm 5 for this gear train in terms of the pitch radii. Use this equation to verify the answers from your Working Model simulation for all three cases.

$$\frac{\omega_4 - \omega_5}{\omega_1 - \omega_5} = -\frac{r_1 \cdot r_3}{r_4 \cdot r_2}$$

Using chain drives and friction belts

Since the motion of chain drives and friction belts is basically the same as it is for spur gears, you can use the Gear tool to simulate these mechanisms as well. You should place the gears for both chain drives and friction belts where you want them in the workspace (for these problems, there will be a space between the two gears), and use the Gear tool to connect the two circles. You will need to use one internal gear and one external gear to model the operation of a chain drive (like a bicycle chain) or friction belt in Working Model. You can then specify the gear ratio as required in the particular problem.

You have completed your examination of the speed ratios of all three cases.

Experiment now if you wish by substituting a square for one of the circles to see that Working Model does not take into account the geometries of the gear objects.

■ **Exit** Working Model.

13.1 The Basics When the automotive industry is not hiring, maybe the bicycle business is good. Use the Polygon tool to create a triangle to simulate the rear portion of a bike frame. Add a round wheel pinned to the frame and a rectangular crank attached with a motor to the frame. Make it a velocity motor with an angular velocity of -400°/s. (Do you know why you are using a negative sign?) Use the Gear tool to connect the wheel to the crank, as shown in Figure E13.1. Make the gear ratio -1 and run the simulation. Why is a negative gear ratio appropriate here? Change the gear ratio to 0.5 and 2 and rerun the simulation for both cases. Figure out whether the gear ratio is the angular velocity of the first object divided by that of the second, or the other way around.

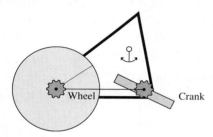

Figure E13.1

13.2 Applying the Basics Figure E13.2 shows a motor, a gear train, and a load. (A *load* is what the system tries to move, for example the prop-water interface in a motor boat.) Circle C has a radius of 0.8 m and is powered by a 10-N-m torque motor. Circle B has a radius of 0.5 m and is connected to circle C with a rigid joint at their centers. Circle A has a radius of 1 m, is geared to circle B, and has a 0.1-N-m-s/rad rotational damper at its center. Run the simulation and see what final speed is attained by circles A and B. Can you use a free-body diagram and equations of motion to compute the system's final speed?

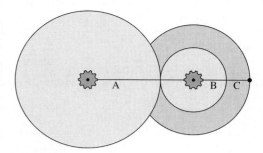

Figure E13.2

13.3 For Fun and Further Challenge This problem examines a pendulum mechanism for a clock. Create two different-sized circles side by side, as shown in Figure E13.3. Pin the left circle to the background with a pin joint and connect the two circles with the Gear tool. Allow Gravity to act on the system. Run the simulation. Monitor the time and rotational velocity (angular velocity) of each circle. Can you create a free-body diagram that shows why the circles move as they do? What is the period of oscillation of the system? (How much time does it take for the system to return to its starting point?) Does the mass of the circles affect the period? Repeat the steps of this simulation, but use a rigid joint to attach the left circle to the background. What difference does this make?

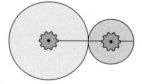

Figure E13.3

13.4 Design and Analysis Open the Design and Analysis simulation from Exercise 12.4. Add a circle pinned to the body to represent the motor. Give it a torque of -120 N-m (the negative sign will cause the motor to rotate clockwise). Attach a gear constraint between the motor and the rear roller with a gear ratio of 0.2. Your workspace should look like Figure E13.4. Run the simulation. Monitor the vertical and horizontal velocity and the angular velocity of the vehicle. You may need to make the ground longer to keep the vehicle from driving off into undefined space. Save this simulation.

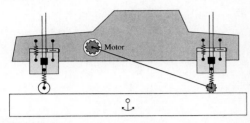

Figure E13.4

Exploring Your Homework This problem uses an internal gear, an external gear, and a link. You can use this geometry to give large gear ratios for power transmission.

13.5 The disk shown in Figure E13.5 rolls on the curved surface. The bar rotates at 10 rad/s in the counterclockwise direction. Determine the velocity of point A.

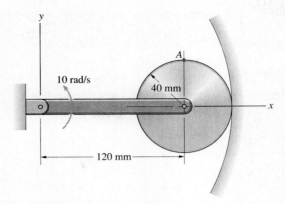

Figure E13.5

Exercise 13.5 is adapted from Bedford and Fowler's Engineering Mechanics: Dynamics (Addison-Wesley, 1995), page 252.

Dynamic Balancing of an Internal Combustion Engine

Objectives

In this tutorial you will learn to:
- Change the location of the mass center
- Define force vectors to illustrate inertia forces
- Copy output meters from one simulation to another

Design principles related to this tutorial

- Dynamic force analysis of slider-crank mechanisms
- Dynamic balancing of reciprocating masses
- Design of V-shape multiple-cylinder engines

Introduction

The transmission of periodic inertia forces to the supports of a reciprocating engine results in undesirable noise and vibrations. In this tutorial you will create Working Model simulations to examine several possibilities for balancing inertia forces in internal combustion engines.

Problem Statement

An internal combustion engine is composed of several slider-crank mechanisms that can be arranged in different ways. Create a Working Model simulation of a single slider-crank mechanism, as shown in Figure 14.1(a), with the centers of mass for the crank and connecting rod located off-center. Put force vectors showing forces acting on all the parts. To the extent that these forces are non-zero, the engine is imbalanced. Add a balancing mass to see whether it improves the balance of this system. Next create a two-cylinder V-shape engine, and analyze its state of balance. The V-shape two-cylinder engine is shown in Figure 14.1(b). The two slider-crank mechanisms share the same crank, and their connecting rods and pistons have the same size and shape. The following information is given for the system:

Speed of the crank (ω) = 1500 rpm

Length of the crank (R) = 3 in.

Length of the connecting rod (L) = 15 in.

Weight of the crank (W_c) = 5 lb

Weight of the connecting rod (W_r) = 15 lb

Weight of the piston (W_p) = 9 lb

Distance from the center of mass of the connecting rod to the crank pin (L_c) = 5 in.

Distance from the center of mass of the crank to the crank pin (R_c) = 1 in.

Figure 14.1

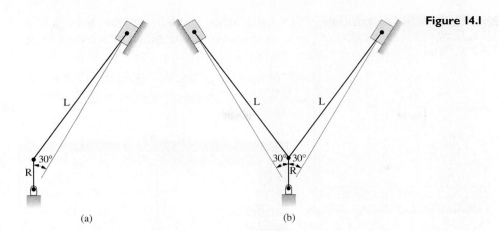

(a)　　　　　　　　(b)

Setting Up the Workspace

- **Open** a new Working Model window with rulers and axes on as before.

- Be sure your simulation is set for **Accurate** mode.

You will set up the workspace in terms of the given units and the size of the slider crank parts.

- Use the Numbers and Units dialog box to change **Unit Systems** to **English (slugs)**.

W Select **More**.

M Select **More Choices**.

- Change the **Distance** pull-down menu to **inches**.

- Change the **Time** pull-down menu to **minutes**.

- Change the **Rotation** pull-down menu to **revolutions**.

- Click **OK**.

- Use the View Size dialog box to change the objects on the screen to **0.15** times actual size.

Creating the Components

You will create and properly position the elements of the slider-crank mechanism, i.e., the crank, connecting rod, piston, and motor.

- Place three **rectangles** with the following dimensions in the workspace: .5 in. × 3 in. (crank), .5 in. × 15 in. (connecting rod), and 1.5 in. × 2.5 in. (piston).

- Locate the left end of the crank at the **origin** of the workspace so the coordinates of its center of mass are (**1.5,0**).

Tip You may need to turn off Grid Snap to create and place the rectangles easily in your workspace. Remember to turn Grid Snap back on when you have finished.

Your workspace should look similar to Figure 14.2.

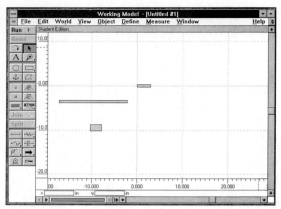

Figure 14.2

- Use the Motor tool, the Help Ribbon, and the Properties window to connect a **motor** to the crank at the origin.

- Give the motor a **velocity** of **1500** rev/min.

  Tip Remember to first locate the motor on the crank; you can then adjust the location of its base point and motor point as needed. If the motor point is not connected to the mass for the crank, delete the motor and try again.

You now need to create a slot on which the piston will reciprocate. This slot will be at an angle of 30° from the vertical, as shown in Figure 14.1(a).

- Choose the **vertical Slot** tool on the tool bar (it pops up from the horizontal Slot icon).

- Click on the **y axis,** away from the origin (otherwise the slot might attach itself to the crank).

For a review of the Slot tool, refer to Tutorial 11. Your workspace should look like Figure 14.3.

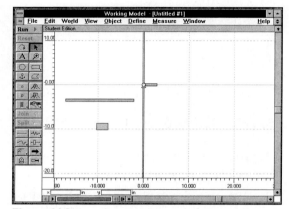

Figure 14.3

- Open the **Properties** window for the slot.

The Properties window should look like Figure 14.4. The slot constraint is Constraint [8] and it has one point associated with it, Point[7], in the box labeled Slot. This is the point about which the slot will rotate if the angle of the slot is changed. (It's the point you picked when you created the slot, so your coordinates may differ.) Because this slot will need to be rotated with respect to the origin, it must be located at the origin. You will use the Properties window to locate Point[7] at the origin.

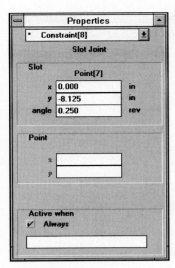

Figure 14.4

- Change the y coordinate of Point[7] to **0** in the slot's Properties window (the x coordinate should already be 0 if you clicked on the y axis).

The point is now located at the origin, but the change is not visible in the workspace. It was helpful to have the rotation units in revolutions for entering the motor speed; when describing the angular position of the crank and slots, it will be more convenient to use degrees to indicate rotation. Because Working Model measures degrees counterclockwise from the positive x-axis, you will set the angle to 60°.

- Use the Numbers & Units dialog box to change the **rotation** from revolutions to **degrees**.

- Set the angle to **60°** for the slot.

Your workspace should look like Figure 14.5; the slot has rotated 60°.

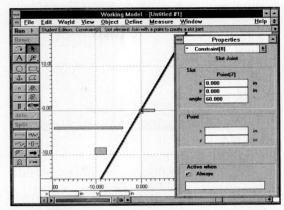

Figure 14.5

- Use the **Properties** window for the piston to set its angle to **60°**.

- Place a **square point** at the center of the piston.

The Properties window for the square point shows its angle relative to the piston. You want this angle to be zero relative to the piston as indicated by the Properties window.

- If necessary, change the angle, ϕ, for the square point to **0**.

Your workspace should resemble Figure 14.6. (A square point appears to have an angle of zero no matter how it is oriented in the workspace, so the icon for the point in your simulation won't change.)

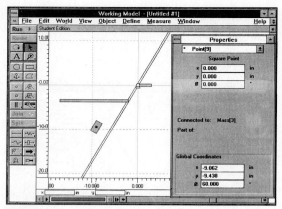

Figure 14.6

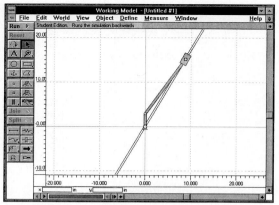

Figure 14.7

- Select the slot and the square point and **join** them.

- **Click** and **drag** on the piston to locate it toward the top of the slot, above the crank, approximately as shown in Figure 14.1(a).

- Place **round points** at the endpoints of the connecting rod and the free end of the crank, and another **round point** at the center of the piston, using the Properties window to locate them correctly.

- **Join** the left end of the connecting rod to the point on the crank.

- **Join** the free end of the connecting rod to the point on the piston.

- Use the Properties window to confirm that the angle of the crank is **90°**.

Your workspace should look like Figure 14.7.

- Close the **Properties** window.

- Select **View** menu, **Lock Points**.

You will now adjust the weights and the locations of the mass centers of the objects, as required by the Problem Statement.

Changing the Location of the Mass Center

Connecting rods and cranks are not rectangles; in reality they have a more complex geometry, and their mass centers are generally not located midway between their endpoints. In this problem, the locations of these mass centers have been defined. You will first display the mass centers for the three objects, and then relocate them for the connecting rod and crank. You can display mass centers using the Appearance window, and move them by means of the Geometry window.

- Select the crank, connecting rod, and piston.
- Open the **Appearance** window.
- Click the **Show center of mass** button.
- Close the **Appearance** window.

The centers of mass (located by default at the mass centers) are now indicated by an **x** on all three objects, as shown in Figure 14.8. You will now relocate the mass centers for the crank and connecting rod, as required in the Problem Statement.

- Click on an empty area of the workspace to deselect all items.
- Select the crank.
- Open the **Geometry** window.

The COM (Center of Mass) field of the Geometry window is set to Auto by default.

This places the center of mass at the geometric center of the body (0,0). The Problem Statement requires that the center of mass of the crank be located 1 in. from the connecting rod pin; this places it .5 in. from the geometric center of the crank.

- Click the button below the **Auto** button in the **COM** field.

The x and y offset directions in the Geometry window are based on the original x, y orientation of the object. Because the original x direction for the crank is now oriented along the y direction, you will use the x offset, but the crank's center of mass will appear to move in the y direction. Positive and negative values for the offset are also based on the original x, y orientation.

- Change the **x offset** to .5 in.

Figure 14.8

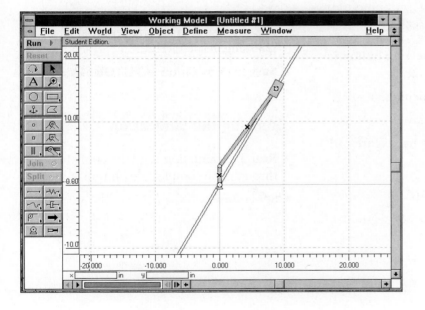

Your workspace should look like Figure 14.9.

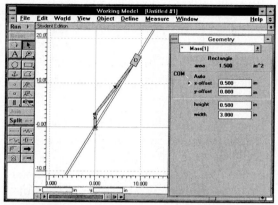

Figure 14.9

As required in the Problem Statement, you will locate the center of mass of the connecting rod 5 in. from the crank pin; i.e., 2.5 in. from the geometric center of the connecting rod.

- Select the connecting rod.
- Click the button below the Auto button in the **COM** field.
- Change the **x offset** to **-2.5** in.
- Close the **Geometry** window.

Your workspace should look like Figure 14.10.

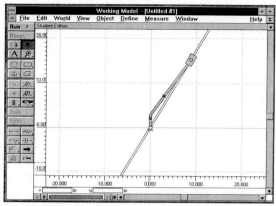

Figure 14.10

The weights of the crank, the connecting rod, and the piston given in the Problem Statement are 5 lb, 15 lb, and 9 lb, respectively; i.e., their masses are ($W/g = W/32.2 \text{ ft/s}^2$) .155, .466, and .280 slug, respectively.

- Use the Properties window to change the masses for the crank, the connecting rod, and the piston to **.155** slug, **.466** slug, and **.280** slug, respectively.
- **Save** the simulation as **TUTOR14A**.

Running the Simulation

- **Run** the simulation until the crank has gone through one complete revolution.
- Select **Reset**.

The motion is slow and jerky. This may occur in high-speed dynamic problems if the default time step used in the Working Model algorithms is too large. You will decrease the time step by one order of magnitude to see if this will improve the simulation. For more detailed information on the Working Model time steps and accuracy, refer to Appendix A.

- Select **World** menu, **Accuracy...**

- In the **Animation Step** field, click the button below Automatic.

- Change the number in the upper box from 1.667e-04 to **1.667e-05**.

- Press **Tab**.

Your dialog box should look like Figure 14.11.

Figure 14.11

- Click **OK**.

- **Run** the simulation until the crank has gone through one complete revolution.

- Select **Reset**.

The simulation now runs more smoothly, because calculations are being performed at smaller time intervals.

Obtaining Results

- Open output **meters** for the **total force** acting on the crank, connecting rod, and piston in **digital** display.

- Eliminate all but the display of the magnitude of the total force, |F|, for each meter.

- Name and position these meters appropriately.

Your workspace should look like Figure 14.12.

Figure 14.12

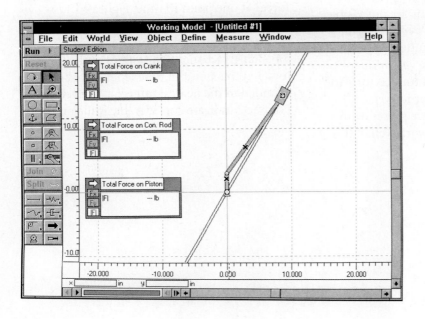

Defining force vectors to illustrate inertia forces

As you have seen, Working Model can display vectors during the motion of your simulation. These vectors can be shown at constraints or at the mass centers of objects. Force vectors displayed at the mass centers of objects illustrate the inertia (imbalance) quantities (i.e., mass × acceleration, according to Newton's Laws) resulting from external loading imposed on the system (in this case, the motor rotation). You will display vectors for the inertia forces for this system. Wherever force vectors appear in the simulation, imbalance occurs.

- Select the crank, the connecting rod, and the piston.
- Select **Define** menu, **Vectors, Total Force**.
- **Run** the simulation until the crank has gone through one complete revolution.
- Select **Reset**.

Note that the vectors are too long, as shown in Figure 14.13; the vector heads are not visible. You must change the force's vector length. (Remember that the value in the Vector Lengths dialog box times the vector quantity equals the length of the resulting vector.) You must decrease the scale factor for vector length, because the vectors are presently too long. A value of 1.0e-6 will work well in this case.

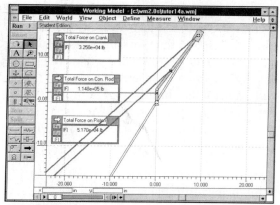

Figure 14.13

- Select **Define** menu, **Vector Lengths...**
- Change the value under Force to **1.0e-6**
- Click **OK**.
- **Run** the simulation until the crank has gone through one complete revolution.
- Select **Reset**.

The vector lengths are now clear throughout the motion. You will examine the inertia forces at one point in time. Figure 14.14 shows the mechanism at the time when the crank, connecting rod, and piston are in line (Frame 37). Table 14.1 shows the values of the total forces at this point in time (they have been rounded to the nearest integer). The vectors at Frame 37 indicate considerable imbalance.

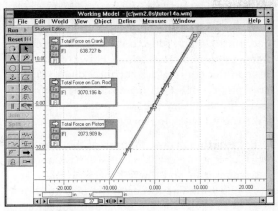

Figure 14.14

Table 14.1 Inertia Forces (lb) at Frame 37

Case	Crank	Connecting Rod	Piston
No balancing	639	3070	2074

Link to Design

The following equations give the inertia forces for the crank, connecting rod, and piston in terms of the variables defined in this problem. F_c is the total inertia force in the crank, F_{rx} and F_{ry} are the x and y components of the inertia forces in the connecting rod, and F_p is the total piston inertia force. Where θ is measured counterclockwise from the position where R and L are in line and the piston is as far as it can get from the crank.

$$F_c = \frac{W_c\left(R - R_c\right)\omega^2}{g}$$

$$F_{rx} = \frac{W_r\omega^2}{g}\left[R\cos(\theta) + L\left(\frac{R}{L}\right)^2\cos(2\theta)\right]$$

$$F_{ry} = \frac{W_r\omega^2}{g}\left[\left(L - L_c\right)\left(\frac{R}{L}\right)\sin(\theta)\right]$$

$$F_p = \frac{W_p\omega^2 R}{g}\left[\cos(\theta) + \left(\frac{R}{L}\right)\cos(2\theta)\right]$$

The second line of Table 14.2 shows the values calculated and rounded to the nearest integer, using these equations for crank angle $\theta = 0.0$, the position shown in Figure 14.14. The values obtained using Working Model are all within 1% of the calculated values. You can use these equations to check other values obtained using Working Model by measuring the angle counterclockwise from its zero position, shown in Figure 14.14.

Adding the balancing mass

Now you will modify the simulation to remove the engine imbalance. A common method for decreasing the engine imbalance is to add a counterweight to the crank, equal to the crank weight; this weight is added at such a distance that the mass center of the crank will move to the crankshaft, i.e., to the origin. You can simulate this situation with Working Model by simply moving the mass center of the crank to the origin; you do not need to double the weight of the crank because, with the mass center of the crank at the origin, the results will be the same for any crank weight.

- Select **Reset**.

- Use the **Geometry** window for the crank to offset the **center of mass** to **-1.5** in.

- **Run** the simulation until the crank has gone through one complete revolution.

- Select **Reset**.

Note that no force vector appears on the crank during the simulation; you have effectively balanced the crank inertia forces. Use the Tape Player to move the simulation to Frame 37, and compare your answers with those in the first line of Table 14.2 and with Figure 14.15.

Table 14.2 Inertia Forces (lb) at Frame 37

Case	Crank	Connecting Rod	Piston
Crank balanced	0	3068	2073
Calculated values	639	3065	2069

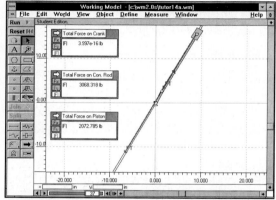

Figure 14.15

| **Link to Design** | There are other ways to try to alleviate imbalance in single-cylinder engines. One |

possibility would be to add a counterweight to the crank, equal to the weight of the crank plus the piston, at a distance *R* from the origin. Another possibility would be to counterbalance the crank weight and a portion of the connecting rod weight at some distance from the crankshaft. Different degrees of balance result from each of these; the balance may be improved in one direction and worsened in another. Counterweights might also be added to the connecting rod, moving the system's center of gravity towards the crankshaft axis.

One focus of the design for balancing such a mechanism is to minimize the *force transmitted to the ground* at the crank's pin-to-ground connection. The forces in the pins connecting the components of the mechanism are also of inter-

est. You can use Working Model to display these forces. For the balanced crank, the inertial load on the crank (due to centrifugal effects) is zero, as demonstrated above. However, the force in the crank pin is *not* zero, due to the transmission of forces from the other components. If you display the pin forces, you can make the following observations:

1. Because the center of mass of the crank is located at the crank pin, it will have zero acceleration, and the net force on the crank will be zero. The forces in the pins at both ends of the crank are equal and opposite.

2. The offset of the lines of action of the forces in the pins at both ends of the crank will change with the angular position of the crank. When the offset is zero, the motor exerts no torque on the crank. At other crank positions, the motor torque is equal to the product of the offset and the magnitude of the pin force.

3. The forces in the pins at the ends of each connecting rod are not equal and opposite, because the acceleration of the center of mass of the connecting rods is never zero. It is composed of both normal and tangential components, which are never simultaneously zero—you can use the Track Connect feature of Working Model to display the path of the centers of mass of the crank and the connecting rod. Note that there is a connection between the path traced and the acceleration of the mass centers of each of these items.

- **Save** this simulation again.

You will now add another slider-crank mechanism to this system, thereby creating a two-cylinder V-shape engine.

- Place two more **rectangles** with the following dimensions in the workspace:

15 in. h × **.5** in. w (connecting rod), and **1.5** in. h × **2.5** in. w (piston).

- Place a **square point** at the center of the new piston.

You want this piston to be oriented as shown in Figure 14.1(b).

- Rotate the new piston to **120**°, using the **Properties** window.

- If necessary, change the angle, φ, for the square point to **0**.

- Place a **vertical slot** along the y axis.

- Use the slot's **Properties** window to move its point to the **origin**.

- Rotate the slot to **120**°, using the **Properties** window.

- **Join** the slot and the square point.

The piston moves to the slot, as shown in Figure 14.16.

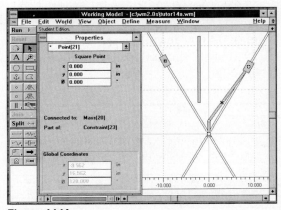

Figure 14.16

- Place **round points** at both endpoints of the new connecting rod.

- Place a **round point** at the top end of the crank.

- Place a **round point** at the center of the new piston.

Tip Use the Help Ribbon when placing points on the connecting rod. If you still have trouble, switch off Grid Snap. Remember to turn it back on when done.

- **Join** the new crank pin and the bottom new connecting rod point.

- **Join** the new piston point and the other new connecting rod point.

Your V-shape two-cylinder engine is assembled as shown in Figure 14.17. You must change the masses and the locations of the objects' centers of mass to be the same as in the other engine.

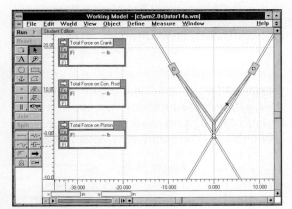

Figure 14.17

- Select the new connecting rod and piston.

- Use the **Appearance** window to turn on **Show center of mass**.

- Select the new connecting rod *only*.

- Use the **Geometry** window to change the COM **y offset** to **-2.5** in.

- Use the **Properties** window to change the mass of the new connecting rod to **.466** slug and the mass of the new piston to **.280** slug.

- **Define vectors** that show **total forces** acting on the new connecting rod and piston.

Items that do not share constraints must be prevented from colliding with each other during the motion. In this case, if you were to try to run the simulation without activating Do Not Collide, the connecting rods would interfere with each other and you would get an error message indicating that objects are overlapping beyond the specified tolerance. You will use the Do Not Collide option to prevent the two connecting rods, and their constraints, from colliding.

- Select all objects.

- Select **Object** menu, **Do Not Collide**.

- **Run** the simulation until the crank has gone through one complete revolution.

The values in the meters at Frame 37, rounded to the nearest integer, are the same as when you last ran the simulation (the crank is still balanced). The sum of all the inertia forces felt at the crank shaft is different, since there are now additional inertia forces resulting from the new piston assembly. You will now examine the total inertia forces in the engine.

- Select **Reset**.

- Open the **Properties** window for the existing piston output meter.

- Maximize the size of the Properties window.

The equation for F_x looks like this: 0.225*(mass[3].mass*14.600*mass[3].a.x *7.056e-06). This formula represents mass × acceleration, *ma*, for Mass[3], the piston. (The constants are the result of the changes you made in the Working Model unit system.) You want to add together four such expressions, for the two pistons (Mass[3] and Mass[19]) and the two connecting rods (Mass[2] and Mass[20]). They should read (depending on the item numbers for your objects):

FxTOT =
0.225*(mass[3].mass*14.600*mass[3].a.x *7.056e-06)+0.225*(mass[2].mass*14.600 *mass[2].a.x*7.056e-06)+0.225*(mass[19]. mass*14.600*mass[19].a.x*7.056e-06)+0.2 25*(mass[20].mass*14.600*mass[20].a.x*7. 056e-06)

Figure 14.18

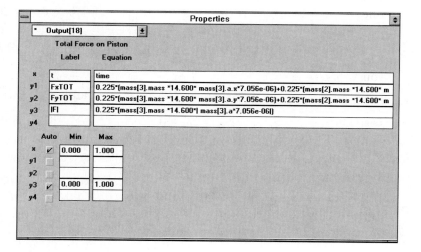

and

FyTOT =
   0.225*(mass[3].mass*14.600*mass[3].a.y
   7.056e-06)+0.225(mass[2].mass*14.600
   *mass[2].a.y*7.056e-06)+0.225*(mass[19].
   mass*14.600*mass[19].a.y*7.056e-06)+0.2
   25*(mass[20].mass*14.600*mass[20].a.y*7.
   056e-06)

- Modify the **Fx** and **Fy** equations so that they
  represent the total inertia forces in the x and
  y directions, respectively, on the two pistons
  and the two connecting rods.

   Tip Use the keyboard commands to copy
   the equation for Mass[3], then paste and
   modify it for each of the other 3 masses.
   Use **W** Ctrl+C to copy and Ctrl+V to
   paste; or **M** Command+C to copy and
   Command+V to paste. Use the left and
   right arrows to move to portions of the
   equation you cannot see on your screen.

If you need help customizing output meters,
refer to Tutorial 8. Your output meter's
Properties window should look something
like Figure 14.18, depending on the IDs for
your objects. The equations are very long and
cannot be seen all at once.

- Change the labels to **FxTOT** and **FyTOT**.

- Clear the other two output meters from
  your workspace.

- Label the revised meter **Total Inertia Force**.

- Display only the values for **FxTOT** and
  FyTOT.

- Use **Save As** to save your simulation as
  TUTOR14B

Your workspace should look like Figure 14.19.

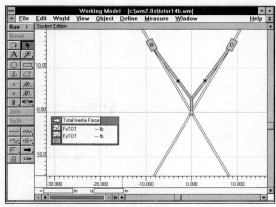

Figure 14.19

- **Run** the simulation until the crank has gone
  through one complete revolution.

- Select **Reset**.

Move the simulation to Frame 10, as shown in
Figure 14.20. Note that even though there are
large forces on both pistons and connecting
rods, the total force due to the pistons and
connecting rods in the y direction is small.
Now look at Frame 20, as shown in Figure
14.21. The x component of the sum of the piston and connecting rod forces is very small,
even though separately the forces are large.
These results are recorded in Table 14.3. The
additional cylinder has changed the net force
considerably (you will soon switch back to the
single-cylinder engine to compare values).
Thus far you have balanced the crank. There
are some systems that you cannot balance
completely. Counterweights might help
reduce the state of imbalance in this system.

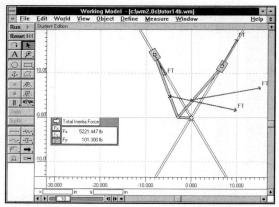

Figure 14.20

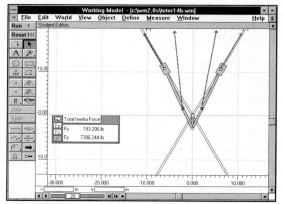

Figure 14.21

Table 14.3 Total Inertia Force (lb)
2-Cylinder V-Shape Engine

	Total x Force	Total y Force
Frame 10	5221	101
Frame 20	193	7396

Copying output meters from one simulation to another

It would be interesting to compare the total inertia forces produced for the single-cylinder

engine with the above results. To do this, you will copy the revised output meter to the first simulation, TUTOR14A.

- **Reset** your simulation.
- Select the output meter.
- Select **Edit** menu, **Copy Data**.
- Select **File, Open...**
- Select **TUTOR14A**.
- Select **OK**.

The single-cylinder simulation opens in the workspace.

- Select **Edit** menu, **Paste**.

The revised meter appears in the workspace.

> **Tip** You can copy any of the objects in your simulation to another file by choosing the object(s), selecting **Copy** from the **Edit** menu, opening the destination file, and selecting **Paste** from the **Edit** menu.

- Locate the new meter as shown in Figure 14.22.

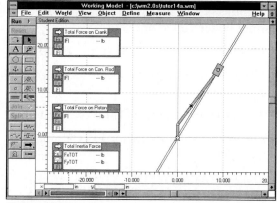

Figure 14.22

- Modify the meter's Properties window so that the formulas include only Mass[2] and Mass[3], which exist in this simulation.

- **Run** the simulation again.

- Select **Reset**.

Use the Tape Player to observe the values of forces at Frames 10 and 20. Compare your results with those in Table 14.4, and Figures 14.23 and 14.24. Notice that, although the second cylinder reduces certain inertia forces, it considerably increases others. You can see how this happens by observing the force vectors and noting the positions at which a particular force component will tend to cancel out a similar component, reducing the imbalance in a certain direction; similarly, there are positions when components will tend to aggregate, increasing the state of imbalance.

Table 14.4 Total Inertia Force (lb) 1-Cylinder Engine

	Total x Force	Total y Force
Frame 10	2691	1255
Frame 20	1123	3712

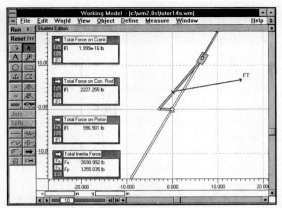

Figure 14.23

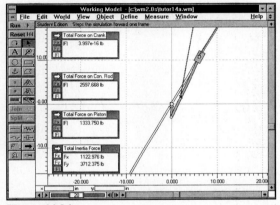

Figure 14.24

You can switch between simulations by using the Window menu and selecting whichever simulation you would like to look at from the list of open files at the bottom of the menu.

- **Save** your simulations and **exit** Working Model.

Tutorial 14 • Exercises

14.1 The Basics Create a slot oriented at 45°, with its slot point at the origin. Add a 1-kg rectangle, also oriented at 45°, as shown in Figure E14.1. Use a round point to pin the rectangle to the slot at its geometric center. Measure all the velocity of the mass and define a total force vector for the mass. Run the simulation under the influence of gravity and standard air resistance and note the force and velocity characteristics of the simulation. Move the center of mass approximately halfway to the upper end of the rectangle and rerun the simulation, monitoring the same variables. What do the results of the imbalance suggest to you?

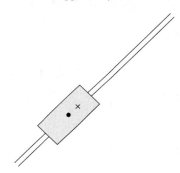

Figure E14.1

14.2 Applying the Basics This exercise will examine the piston-cylinder arrangement of the tutorial in a different way. Create a 10°-rotated slot and attach a 0.2-m piston to the slot with a square point. Over the slot, attach a 0.3-m diameter circular crank to the background with a pin joint, as shown in Figure E14.2. Add a 0.5-m connecting rod and join it with pin joints to both the piston and the crank near its circumference. Make a force act on the piston to represent the effects of combustion. This force should only be active when the piston is moving to the right (use the Active when feature). Run the simulation, monitoring the

crank's angular velocity and the inertia force of the piston. Move the center of mass of the connecting rod and run the simulation again, noting the differences.

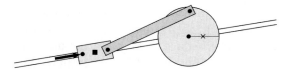

Figure E14.2

14.3 For Fun and Further Challenge You are in the business of precision drafting equipment and you want to create mechanisms for drawing specific shapes. Create two slots, oriented at 30° relative to each other. Add a rectangle with two points on the ends and join them to the slots, as shown in Figure E14.3. Dimensions are not critical, as long as it is easy to see and work with the mechanism. Position your rectangle at an initial angle other than zero so gravity will set the system in motion. Track the center of mass of the rectangle, and run the simulation. What shape appears? Change the position of the center of mass or the relative angle between the slots and run the simulation again. How do these variables change the shapes drawn? What can you say about various forms of energy and energy conservation during the course of the simulation?

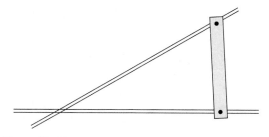

Figure E14.3

14.4 Design and Analysis Open the design simulation from the last tutorial, Exercise 13.4. Add an uneven road under the vehicle with the Polygon tool and display the body's center of mass, as shown in Figure E14.4. Monitor the body's acceleration and the net force on the body as the vehicle moves down the road. You may want to change the tire size to be more realistic. What are the effects of changing the body's center of mass position? Why does center of mass change in real life? Save this simulation.

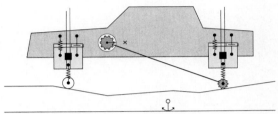

Figure E14.4

Exploring Your Homework A rotated slot and a spring are necessary when modeling this problem in Working Model. Energy methods can verify Working Model's solution.

14.5 A spring-powered mortar is used to launch 10-lb packages of fireworks into the air as shown in Figure E14.5. Each package starts from rest with the spring compressed to a length of 6 in.; the free length of the spring is 30 in. If the spring constant is $k = 1300$ lb/ft, what is the magnitude of the velocity of the package as it leaves the mortar?

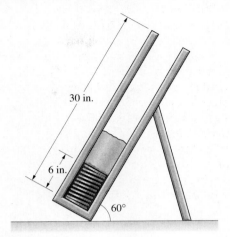

Figure 14.5

Exercise 14.5 is adapted from Bedford and Fowler's Engineering Mechanics: Dynamics *(Addison-Wesley, 1995), page 159.*

Design of a Batting Machine

Objectives

In this tutorial you will learn to:
- Use the Separator tool
- Create torques

Design principles related to this tutorial

- Product design and development

In this final tutorial you will use Working Model to develop a design concept and quickly bring it to the point where a prototype can be built. This process demonstrates the usefulness of Working Model for conceptualizing design and visualizing dynamic behavior prior to building physical models. You will use two Working Model tools, separators and torques, that you have not yet used.

Problem Statement

As a design engineer and the elder sibling of a Little Leaguer, you have volunteered to build a machine that will hit softballs to the infield and outfield for the purpose of helping the players to practice their fielding techniques. You already have a concept for such a device, sketched in Figure 15.1. Now create a Working Model simulation for development purposes. Your concept includes the following:

1. Softballs (radius $R = .075$ m) will be fed by a chute that delivers them at a constant velocity with a constant distance between them.

2. A bat (length $L = 1$ m) will hit the balls. In order to maintain safety and conserve space, the bat will be prevented from rotating more than 90°; its motion will be governed by a torquing device that is driven by a linkage attached to a motor; the bat will hit the ball, and then rotate back to its starting position.

When your Working Model simulation is complete, examine the contact forces produced between the bat and the ball so that you can develop guidelines for the selection of the bat material. The coach should be able to change the ball-feed velocity and the distance between the balls in order to deliver the ball at different speeds to infield and outfield positions. There will be trial and error involved in the development of the batting machine—not every design approach will work; Working Model provides you with the opportunity to try various scenarios *before* building a prototype.

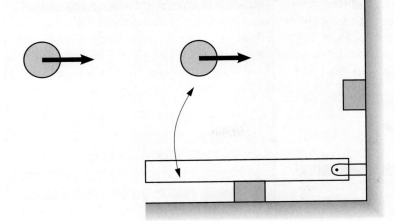

Figure 15.1

Setting Up the Workspace

- **Open** a new, untitled Working Model window with the rulers, grid, and axes as before.

- Use Working Model's default settings for Numbers and Units (SI (degrees)) and Accuracy (Accurate mode).

Initially, in order to observe the details of the design, you will change the workspace so that the 1-m bat takes up about half the screen; i.e., the screen should appear to be approximately 2 m wide, or one-fourth the size it now appears (the x axis default width is approximately 8 m).

- Select **View** menu, **View Size...**

The View Size dialog box shows the default object size as 0.021 times actual size, and the scale (or window width) as about 8 m. Your number may be slightly different, depending on the size of your monitor.

Since you want objects to appear approximately 4 times larger than they would by default, you will change the scale (or window width) to 2 m.

- Change the value in the lower text box to **2** m.

- Click **OK**.

Since you are creating a bird's eye view of the system, you do not want gravity acting on the simulation.

- Use the Gravity dialog box to set **Gravity** to **None**.

Creating the Components

You will create and assemble the bat, the housing for the machine, the torque, and four softballs.

- Create a **rectangle** 1 m wide × 0.1 m high (the bat) positioned with the center of its right endpoint at the origin.

Your bat should look like Figure 15.2.

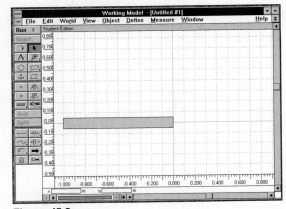

Figure 15.2

Creating the Components ■ 243

- Using the **Polygon** tool, create a housing for the batting machine. The coordinates of the polygon's vertices are **(-1,-.1)**, **(-1,-.2)**, **(.2,-.2)**, **(.2,1)**, **(.1,1)**, and **(.1,-.1)**. (Remember to double-click at your last point to close the Polygon.)

- **Anchor** the housing.

Your workspace should look like Figure 15.3.

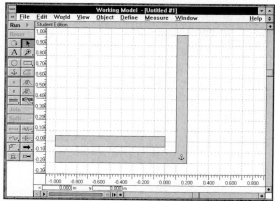

Figure 15.3

You will use a pin joint to represent the turning joint on the bat.

- Join the bat to the background precisely at the origin with a **pin joint**.

Your workspace should look like Figure 15.4.

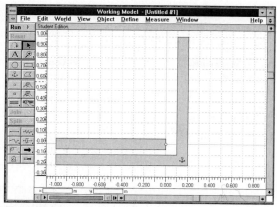

Figure 15.4

Using the Separator Tool

The *Separator* tool pops up from the Rope icon on the tool bar. Separators prevent objects from moving closer together than a specified distance. You will use separators in this problem to prevent the bat from rotating more than a total of 90°. A separator has no effect when the objects it is connected to move away from each other; the separator just continues to open. You can attach separators between one mass object and the background, or between two mass objects (the endpoints of the separator are the attachment points). Separators can be thought of as the opposite of ropes. Ropes prevent their endpoints from being farther than a certain distance from one another; separators prevent their endpoints from being *closer* than a certain distance to one another. You will create two separators in your model: one to keep the bat from the bottom of the housing, and one to keep the bat from the side of the housing.

- Select the bat.

A small circle appears at its mass center.

- Click and hold down on the **Rope** icon on the tool bar.

- Select the **Separator** option (to the right of the Rope icon).

- Click and drag from the mass center of the bat, directly down to the vertical midpoint of the horizontal portion of the housing.

Your screen should look like Figure 15.5.

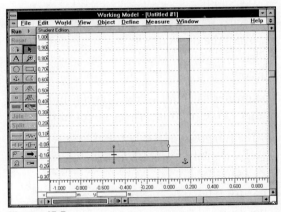

Figure 15.5

You have created a separator that will prevent the bat from striking the housing when it is turning counterclockwise. You will now create a similar separator that will prevent the bat from striking the housing when it is turning clockwise. The separators model the bumpers on the housing in Figure 15.1.

- Select the bat again.

- Use the **Properties** window to set angle ϕ to **-90°**.

Your workspace should look like Figure 15.6.

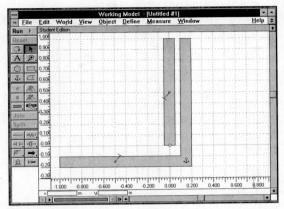

Figure 15.6

- Click on the **Separator** icon again.

- Click and drag from the mass center of the bat directly to the right, to the horizontal midpoint of the right side of the housing.

Your workspace should look like Figure 15.7.

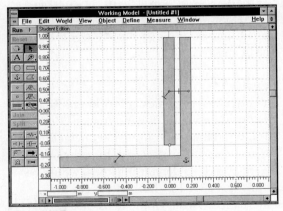

Figure 15.7

- Open the **Properties** window for the bat.

Charge, density, and moment of inertia

The Properties window for the bat includes the following fields: Charge, Density, and Moment of Inertia.

Charge and electrostatics

Charge dictates how a mass behaves in an electrostatic field. Mass objects are given an initial charge large enough to produce movement between human scale (1.0-m) objects. Charges only affect the simulation when the Electrostatics feature is turned on. You can turn electrostatics on by selecting Electrostatics from the World menu.

Each mass object in Working Model has a charge. By default, each object has a positive charge of 1.0×10^{-4} Coulombs. This charge is enough to produce interesting results between objects in the default physics workspace, although it really is an extreme amount of charge. You will need to set the charge values of various mass objects to values other than 0 to see the effects of charge.

Electrostatics works as if all of the charge of a mass object were concentrated at its center of mass. Charge is not distributed over the surface of a mass object.

Density

Initially, all rigid bodies in Working Model are considered to be one millimeter (1 mm) thick, no matter what unit system you are in. For example, if you drew a 1-foot-by-1-foot square object, its default thickness would be 1 mm, or 3.28×10^{-3} ft. If you chose the material to be steel, whose density is about 500 lb/ft^3, the weight of the object would be shown as $500 \times 3.28 \times 10^{-3}$, or 1.639 lb.

All objects are initially given a density of 1.0 g/cm^3 (equal to the density of water) no matter what unit system you are in.

Larger objects are initially heavier than small objects because they are both given the same density.

You can view the density of any mass object in the Properties window. You indirectly change a mass object's density whenever you specify a new value for its mass. The only way to directly change a mass object's density is through choosing a material.

Moment of inertia

By default, mass objects are assigned moments of inertia because Working Model assumes that they are planar and have a uniform mass distribution.

You can adjust the moment of inertia to that of a shell or spherical weight distribution by choosing the desired moment from the pull-down Moment menu in the Properties window.

The numerical value of the object's moment changes to reflect the new moment of inertia.

For the purposes of this tutorial, you will leave Charge, Density, and Moment of Inertia set to their defaults.

- Return the bat to its original, horizontal position.

- Select the first separator.

The Properties window for the separator, shown in Figure 15.8, shows the length, current length, and elasticity for the separator; at the bottom of the Properties window is the Active when field, which works the same way it did for forces. (For a review of Active when, refer to Tutorial 8.) The length is the actual length of the separator when the endpoints are in their closed position. The current length is the current amount of separation between the endpoints. The elasticity determines the difference between the velocities of the two bodies after a collision, i.e., how elastic a collision will be. The default value is 0, which means that the collision will be perfectly plastic and the two bodies will remain in contact unless some other constraint acts to separate them. You will leave the default value for now, and assume that you will use an energy-absorbent bumper when you build the actual machine. You want the endpoints of the constraint to be at the exact centers of the bat and the horizontal portion of the housing, and the current length to be equal to the length for both the vertical and horizontal positions of the bat, so you will set them equal now. In Figure 15.8, the length and the current length are both 0.150 m.

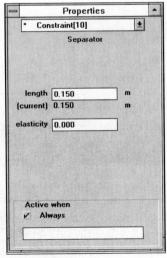

Figure 15.8

For the first separator, the endpoints must be precisely located on the bat and the housing.

The endpoint on the bat should have relative coordinates of (0,0) relative to the bat, and the endpoint on the housing should have global coordinates of (-.5,-.15).

- Use the pull-down menu in the **Properties** window to identify the endpoints of the first separator.

- If the endpoints are not located properly, use the **Split** tool on the separator, relocate the endpoints, and then use the **Join** tool to reconnect them.

- Use the **Properties** window, if necessary, to change the **length** of the separator to **.15** m.

- Move the bat to the vertical position.

- Select the second separator.

For the second separator, the endpoint on the bat should have local coordinates of (0,0) relative to the bat, and the endpoint on the housing should have global coordinates of (.15,.5). Compare these values to the values you entered above. If the endpoints are not located properly, move them as instructed for the first separator.

- Identify the endpoints of the second separator.

- If the endpoints are not located properly, use **Split,** then **Join** to relocate and reconnect them.

- If necessary, change the **length** of the second separator to be equal to that of the first separator, **.15** m.

- Move the bat back to its horizontal position.

- Close the **Properties** window.

Creating Torques

The *Torque* tool pops up from the Force icon on the tool bar. Torques apply a turning force to an object. The Properties window for the torque is used to set its magnitude.

- Click and hold down on the **Force** icon on the tool bar.

- Select the **Torque** icon (an arrow curved around a dot, to the right of the Force icon).

- Click to the left of the pin joint on the bat.

Your workspace should look like Figure 15.9.

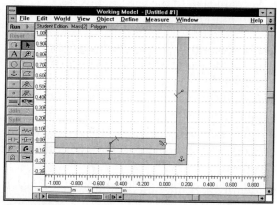

Figure 15.9

Be careful that the torque is applied to the bat and not to the background.

- Open the **Properties** window for the torque.

- Locate the **base point** for the torque precisely at the pin joint, at the relative coordinates **(.5,0)**.

You have now created a torque and located it at the pin joint. (The location of the torque need not be at the joint. The torque will act at the pivot point no matter where it is placed on the mass.) You want the torque to cause the

bat to move clockwise to hit the ball, and then counterclockwise to return to its original position. For the purposes of this tutorial, you will model the torque as a sinusoidal function. You want the torque to be acting clockwise when it hits the balls; clockwise is negative for Working Model, so your function must initially be negative. You will insert the sinusoidal function in the Properties window for the torque control. To figure out what function to use, you must run the simulation and see how long it takes for a torque to cause the bat to rotate clockwise 90°. You will first insert an arbitrary value of -10 N-m for the torque.

- Change the **value** for the torque to **-10** N-m.

- Close the **Properties** window.

- Open a **time** output **meter** in **digital** display.

Running the Simulation

- **Run** the simulation until the bat hits the separator on its right side.

- Select **Reset**.

- Use the **Tape Player** and the time output meter to determine approximately how long it took to hit.

Figure 15.10 shows that it took approximately 0.14 s to hit the separator; i.e., it went half of what you want its complete cycle of motion to be in 0.14 s. The sinusoidal function should be of the form -10sin(ωt), where ωt = ≠, for one half of the cycle. Thus, if you would like it to return in the same amount of time, ω = ≠/0.14, your torque value should be set equal to -10sin(≠t/0.14).

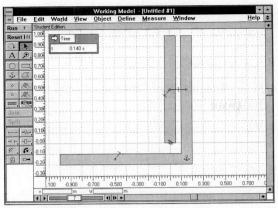

Figure 15.10

- **Reset** the simulation.
- Use the Properties window to change the **value** for the torque to **-10*sin(Pi*t/.14)**

Your Properties window should look like Figure 15.11.

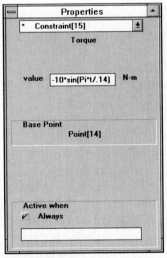

Figure 15.11

You are now starting the detailed design phase of the simulation. You need to place the balls in the workspace so that the bat will hit them

properly. You will use a combination of trial and error and dynamics principles to accomplish this. The initial location of the first ball given below was obtained by trial and error, and the initial velocity was arbitrary. You will see that once the first ball has been successfully placed (so that the bat hits it properly), the approximate location of the others can be obtained theoretically.

- Create a **circle** of radius **.075** m (a softball) and place it at **x = -1, y = .5**.
- Give the softball an **initial velocity** in the x direction of **4** m/s.

Your workspace should look like Figure 15.12 when the ball is selected.

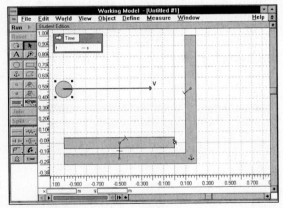

Figure 15.12

- **Run** the simulation to make sure that the bat hits the ball.
- Select **Reset**.

Obtaining Results

Because the time of impact during the torquing cycle and the location of impact on the bat will vary, you would not expect the velocities, directions, and contact forces to be identical for each ball. You can, however, calculate an approximate spacing for the balls. Since the bat motion will recycle (i.e., swing and return) every 0.28 s, and the balls will initially be moving at 4 m/s, they must be spaced approximately 0.28 x 4 = 1.12 m apart. As you work on this section, you may need to use the Zoom tool and scroll bars to more easily observe the behavior of your simulation.

- **Duplicate** the ball three times (its initial velocity will also be duplicated).

- Place the new balls at **1.12**-m intervals in the negative x direction; i.e., the second, third, and fourth balls will be centered at **(-2.12,.5)**, **(-3.24,.5)**, and **(-4.36,.5)**, respectively.

You can see three of the four balls by scrolling the housing to the right side of the workspace and using the Zoom out tool. Refer to Figure 15.13.

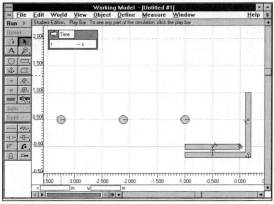

Figure 15.13

As required in the Problem Statement, you will examine the variation in contact forces as the balls hit the bat.

- Select the bat.

- Select **Define** menu, **Vectors, Contact Force**.

- **Run** the simulation until the first impact occurs.

- Select **Reset**.

You can barely see the force vector. You will change the force vector's length definition to a higher value to make the vector's length appear greater.

- Use the Vector Lengths dialog box to change the force vector's length value to **1**.

- **Run** the simulation until all the balls have passed through the batting machine.

- Select **Reset**.

Contact force vectors may not appear for all the balls. This is a sign that Working Model accuracy may not be sufficient; i.e., the time steps are too large to include the instant of impact. You can change the accuracy of the simulation in the Accuracy dialog box by selecting the **W** More option or **M** More Choices option, and changing the Integration time step in the lower left of the dialog box to **Locked** and setting the value to **0.002** s, as shown in Figure 15.14. Running the simulation with these options selected may produce the contact force vector display for each ball. Figure 15.15 shows the contact force vector for ball 1. For more information on accuracy and locked integration time steps, see Appendix A.

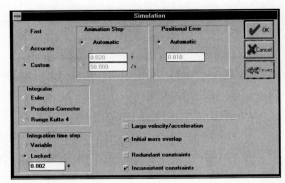

Figure 15.14

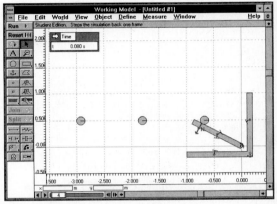

Figure 15.15

Creating contact force meters is the only way to be sure of the point of contact of each ball with the bat.

- Selecting the bat and one ball at a time, create and label a **contact force meter** in **graph** display for each ball *and the bat*. (Contact force meters measure only a pair of objects.)

- **Run** your simulation.

- At the frame at which each ball makes contact with the bat, switch the meters to **digital** display to determine what the scale of the meter should be for easiest viewing.

- Use the meters' Properties windows to change the **scale** of each meter.

Your workspace will look like Figure 15.16. You can try setting different values for the accuracy of the simulation, and for the velocity and elasticity of, and the distance between the balls to see how your results change.

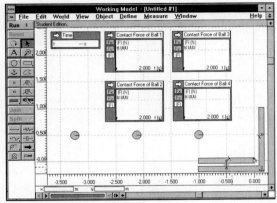

Figure 15.16

Link to Design

The batting machine can operate with different ball spacings and velocities. Try changing the ball velocity, repositioning the first ball, and calculating the required spacing for the other balls. The final machine should be built so that the operator will be able to adjust the spacing and velocities in order to vary the direction, distance, and speed of the balls after they contact the bat. As an additional exercise, you may wish to design the mechanisms that drive the bat and feed the balls. A crank-rocker mechanism (such as the four-bar linkage analyzed in Tutorial 11), driven by a motor, could be used to obtain the rocking motion needed to drive the bat. It should be kept in mind that, since this is a two-dimensional simulation, the height to which the balls will fly has not been dealt with, nor has the projectile angle of the balls from the vertical. These will depend on the relative vertical positions of the ball and the bat as they come into contact, and cannot be handled by the present simulation.

You have completed a design for the batting machine. You may now exit Working Model.

15.1 The Basics You must create a preliminary design of a rotational oscillation test setup. Attach a 1.6-m diameter, 2-kg circular disk to the background with a 10-N-m rotational spring. Add a 2-m bar, rigidly attached to the disk. You may need to split the disk from the spring temporarily to attach the bar. Use separators to keep the disk from rotating more than 30° in either direction, as shown in Figure E15.1. Complete the system with a torque on the disk as a function of time, 1*sin(2*t) N-m. Make the torque active for only the first six seconds of the simulation. Turn off Gravity and run the simulation for 8 seconds. Monitor the angular velocity of the disk and see if you can understand its behavior. Save the tape recording of the simulation.

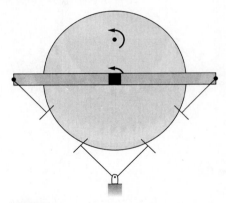

Figure E15.1

15.2 Applying the Basics In the Automation Laboratory, you are designing an automatic door control for the "Factory of the Future." Mobile robots travel the factory and the doors need to sense a robot's presence and motion to open in time to avoid collision. (The robot probably has collision avoidance programming as well.) Set Gravity to None. Create a 0.1 x 1.0-m rectangle with a mass of 10 kg. Use the rotational spring which pops up from the Spring icon on the tool bar to pin the side of the door to the background.

Add a separator so the door cannot open more than 90°. Refer to Figure E15.2. Adjust the spring constant so that when the door is displaced and released, it has an oscillation period of about 2 seconds. Use a rotational position meter to determine this. Add a rotational damper at the same location as the spring. Adjust its constant so that when the door is released from the open position, it closes after only a little swinging back and forth. Add a rectangle representing the robot with a velocity to the right of 1 m/s.

Once the manual door behaves well, add a torque to the same hinge as the spring and damper, so that the torque is a function of how close the robot is to the door. Use a conditional statement so the torque increases as the robot gets closer to the door, but goes to 0 once the robot is safely through the door. Monitor any interesting variables, such as positions and torques. Does the robot make it through unscathed?

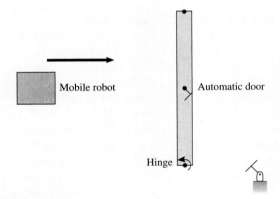

Figure 15.2

15.3 For Fun and Further Challenge You are examining engine motion dynamics during vehicle acceleration. Model the engine as a 100-kg rectangle under the influence of gravity. Support the engine with three springs, two for vertical motion and one for horizontal motion. Determine a spring constant so the engine does not compress the springs more than .020 m due to gravity, and set all three springs to this value. Add dampers, as shown in Figure E15.3, and run the simulation, monitoring vertical motion. The engine mass will settle under the gravitational force and oscillate until the dampers stop the motion. Adjust the damper constants so that about 2 cycles of oscillation are necessary before the motion stops. Now add a torque of 400 N-m to the engine, active for the first second of the simulation. This will simulate the acceleration of the engine. Monitor the angular position of the engine. What happens during acceleration? What happens when the torque has stopped? Can you relate the results to your own driving experience?

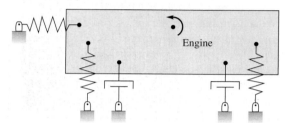

Figure E15.3

15.4 Analysis and Design   Open the design simulation from Exercise 10.4. Add separators to limit the wheel strut (the small rectangles that support the wheels) travel to 8°. This models the "bump stops" of real vehicles, where the wheel travel is constrained in compression by hard rubber stops, as shown in Figure E15.4. Move the center of mass 0.1 m forward and change the motor's torque to 200 N-m. Run the simulation, monitoring the vehicle body's velocities and the separator travel. Rerun the simulation with a motor torque of 400 N-m. Examine the effects of moving the center of mass further forward.

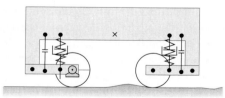

Figure E15.4

Exploring Your Homework   The following problems accelerate a planetary gear set. Position and place the gears carefully and apply the couple with the torque tool instead of a motor.

15.5   The ring gear in Figure E15.5 is fixed. The mass and mass moment of inertia of the sun gear are $m_S = 22$ slugs, $I_S = 4400$ slug-ft^2, respectively. The mass and mass moment of inertia of each planet gear are $m_P = 2.7$ slugs, $I_P = 65$ slug-ft^2, respectively. If a couple $M = 600$ ft-lb is applied to the sun gear, what is the resulting angular acceleration of the planet gears, and what tangential force is exerted on the sun gear by each planet gear?

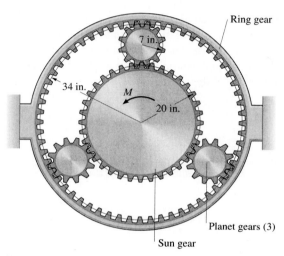

Figure E15.5

15.6   If the system in Exercise 15.5 starts from rest, what constant couple M exerted on the sun gear will cause it to accelerate to 120 rpm in 1 minute?

Exercises 15.5 and 15.6 are adapted from Bedford and Fowler's Engineering Mechanics: Dynamics *(Addison-Wesley, 1995), page 339.*

■ PART THREE ■

Appendixes

Technical Information

This appendix provides you with technical information and describes the inner workings of Working Model. It covers the following subjects:

- Memory requirements
- Running out of memory
- Optimizing speed performance
- Numerical methods
- Time step and performance
- Controlling the simulation parameters
- Integrators
- Animation step
- Integration time step
- Positional error
- Warnings
- Simulating collisions
- Simulation accuracy
- Tips for using Working Model
- Troubleshooting

Memory Requirements

W Windows RAM

Working Model requires 8 MB of RAM, and we recommend using 16 MB. The application can run on 4 MB with virtual memory turned on, but performance will be severely affected.

M Macintosh RAM

We recommend the following MultiFinder or System 7 application memory sizes. Check these (or change the amount of memory allocated to Working Model) by highlighting the Working Model SE icon in the Finder, then choosing Get Info from the File menu. The memory size is determined by the value in the Preferred Size box.

Black and white	3.0	megabytes
4 colors	3.0	megabytes
16 colors	3.5	megabytes
256 colors	4.0	megabytes

The preceding numbers show the requirements for opening one Working Model document at a time. If you want to open two documents at the same time in 256 colors, for example, you need about 4.5 MB of free memory.

Simulation speed depends on the color mode selected. For optimum simulation speeds, use black-and-white or 16-color mode. To change the color options used on your Macintosh, select Monitors in the Control Panel.

Running Out of Memory

Occasionally Working Model may have insufficient system memory (RAM) at its disposal, and it warns you accordingly.

W When running under Windows, you can turn on virtual memory to increase the amount of memory available to Working Model (and the other Windows applications). To do so, run the 386 Enhanced application in the Windows Control Panel. There will be a performance tradeoff. See also Optimizing Speed Performance.

M On the Macintosh, you can increase the amount of memory available to Working Model by changing the application's memory size in the Preferred Size field of the application's Get Info dialog box.

Running simulations unattended

When you create a complex model, you may need to run Working Model for a long time to obtain the simulation results (overnight, for example). If you decide to run the simulation unattended, you should make sure that Working Model can continue its computations uninterrupted. Specifically:

- The memory space allocated to the application should be sufficient to run the simulation and store its results.

- Warning dialog boxes should be disabled so that Working Model can continue computations uninterrupted.

To disable Model-related warnings, select the Accuracy dialog box from the World menu. Click the More button. Remove all check marks from the Warnings check boxes.

Optimizing Speed Performance

The two main factors that affect the performance of Working Model, besides the performance of the computer itself, are available memory (RAM) and monitor color settings.

If you are using full-page or even larger size monitors, you may find that the animation speed drops below acceptable performance. You can increase animation speed by reducing the size of the document window.

W Windows performance

Lack of sufficient RAM causes frequent disk access, which negatively affects performance. This happens because there isn't enough memory to store Working Model and other applications (including Windows) and the operating system needs to swap part of Working Model to disk. To avoid this, quit other applications so that memory is available to Working Model (Windows 3.1 uses at least 800 KB of memory, and Windows for Workgroups uses more than 1 MB). To avoid disk swapping altogether, run the 386 Enhanced application in the Windows Control Panel and turn off virtual memory. Then the use of memory will be limited to installed RAM; the hard disk will not be used to store extra data. This setting obviously limits the number and size of applications you will be able to run.

Simulation speed is also slightly affected by the color mode selected. For optimum simulation speeds, do not use more than 256 colors. To change the colors used in your Windows system, choose Windows Setup from the Main window.

M Macintosh performance

Simulation speed is dependent on the color mode selected. For optimum simulation speeds, use black-and-white or 16-color mode, if available.

If virtual memory is activated, Working Model uses the hard disk for memory transactions, which can slow performance. Select Memory in the Control Panel to deactivate virtual memory (you have to restart your Macintosh to see the result of your change).

Making your simulation run faster

A small change in the model may require a lot less computational effort for Working Model and may speed up your simulation while maintaining sufficient accuracy. Shown below are the list of modifications you may consider applying to your model for faster simulation runs.

- Use the Fast simulation method, and set the time step to the largest value that allows a stable simulation and acceptable accuracy.

- Reduce the number of objects that are in contact. Make sure to use the Do Not Collide option on the Object menu with all groups of objects that do not need to collide. This modification allows Working Model to bypass many collision tests. To visually check for contacts, display collision force vectors by selecting all objects and choosing Define menu, Vectors, Contact Force.

- Set the frictional coefficients of objects in contact to 0.0 if friction is not needed in your simulation.

- Use rigid joints to build complex objects. Using two pin joints to lock objects together introduces extra simulation overhead and redundant constraints.

- Use rods instead of ropes wherever possible.

- Use rods instead of pinned mass objects wherever possible. A truss constructed of small mass objects connected by rods simulates more quickly than a truss constructed of pinned rectangles.

- Make your window size smaller. A smaller window size requires less graphics processing time.

Numerical Methods

Working Model solves problems using numerical methods. A problem is time-discretized so that Working Model can compute motion and forces, while making sure that all the constraints are satisfied. With its systematic approach, Working Model can simulate a wide variety of problems.

Numerical methods is a huge area of ongoing research in science and engineering. Vast amounts of literature are available on this topic. Interested readers are strongly encouraged to refer to more advanced literature.

In describing numerical methods, let's take the example of a ball traveling in projectile motion. For simplicity, consider gravity to be the only force acting on the ball.

Analytical method

An analytical solution for the x and y position of the mass center of a ball seen in most elementary physics texts is:

$$x = x_0 + v_0 t$$

$$y = y_0 + v_{y_0} t + \frac{1}{2} g t^2$$

where $g = -9.81$ m/sec^2.

With these formulas, one can find the position of the ball at any moment in time by simply plugging in the correct initial values ($x_0, y_0, v_{x_0}, v_{y_0}$).

If projectile motion were the only type of problem Working Model needed to solve, this would also be an acceptable way to proceed on the computer.

However, for most physical problems, it is impossible to find exact solutions like those for projectile motion. For example, no analytical solution exists for the equations of motion for three particles (or stars) all acting under gravitational forces among themselves.

Numerical integration

Instead of analytical methods, Working Model uses numerical methods to allow the solution of the motion of mechanical systems that are governed by differential equations arising from mechanics principles. In 2D, these principles can be expressed in simple equations as:

$$F = ma$$

(Force = mass * acceleration)

$$T = I\alpha$$

(Torque = moment * angular acceleration)

$$a = \frac{dv}{dt}$$

(Instantaneous acceleration = time derivative of v)

$$v = \frac{dx}{dt}$$

(Instantaneous velocity = time derivative of x)

$$\alpha = \frac{d\omega}{dt}$$

(Instantaneous angular acceleration = time derivative of ω)

Working Model uses these equations in its solution of dynamics problems. The solution is carried out by a process known as *numerical integration*.

For example, integrating both sides of the equation $a = \frac{dv}{dt}$ results in $v = \int a\,dt + v_0$, which may be approximated as $v = a \cdot dt + v_0$. The accuracy of this approximation improves with a smaller Δt, called a time step (see Time Step and Performance on page 261 for more details). The heart of numerical integration lies in approximating the problem by subdividing the problem into small, discrete time steps, and incrementally computing the result at each time step.

More specifically, Working Model finds the current acceleration of an object, and uses this acceleration to compute a velocity (and position) one time step later. Further, Working Model incorporates a scheme to check and correct its prediction. This process is then used again to find a new velocity and position.

For example, numerical integration for linear motion proceeds as follows:

1. Now at $t = 0$. Calculate a using force equations.
2. Use a to calculate the new velocity v at $t = \Delta t$.

$$v_{t=\Delta t} = a \cdot \Delta t + v_{t=0} \text{ (approximating } v = \int a\,dt + v_0)$$

3. Use v to calculate the new position x at $t = \Delta t$.

There are many methods for numerically integrating acceleration to calculate new velocity and position terms at a later time. The method shown above is known as Euler integration, one of the simplest numerical integrators available. Working Model features two other, more accurate numerical integrators. They are:

Predictor-Corrector

4th-order Runge-Kutta

See Integrators on page 262 for more information on numerical integration.

Time Step and Performance

The size of a time step is a critical parameter in most numerical integrators, and it affects their speed and accuracy significantly. As you might imagine, using a smaller time step results in more accurate predictions for the velocities and positions that are generated as the integrator steps along. In general, a smaller time step produces more accurate simulation results. However, smaller time steps result in more subdivisions per unit time, thus requiring more computational effort per given time period than a larger step size would.

Choosing a proper time step

Choosing a time step for a particular problem can be very difficult. To assist you in your selection of a time step, remember the following rules of thumb:

Small time step = improved accuracy

Large time step = improved computing speed

You do not always have to choose an extremely small time step just because you are concerned about accuracy. More often than not, a reasonably small step size produces a simulation result with sufficient accuracy.

For example, you might begin simulating a problem with a large time step, so you can quickly obtain rough ideas about the model. When you want precise details, you can let Working Model run a long simulation with a smaller time step to verify the accuracy of the model.

Fortunately, Working Model does a good job of shielding you from needing to pick precise time steps, unless you decide to choose them yourself. Working Model has facilities for automatically choosing appropriate time steps and monitoring simulation errors of various types (see Integration Time Step for more details).

For example, an ideal time step should be variable during the course of a simulation, adapting to the complexity of the problem in order to get the most out of the speed-accuracy tradeoff discussed previously. If you are simulating a comet orbiting the sun, you can obtain good results quickly by using a smaller time step when the comet is close to the sun and traveling at a high velocity, and a larger time step when the comet is out past Saturn, where it travels relatively slowly. If you are simulating an automobile collision, you could quickly approach the critical part of the experiment with a large time step while the car is nearing an obstacle, and subsequently use a smaller time step during the collision. If the small time step was used the whole time, the computer might unnecessarily spend a long time computing an acceptable solution to the problem. The default option in Working Model automatically adjusts the time step size during the course of a simulation. See Integration Time Step for details.

Controlling the Simulation Parameters

There are two ways of using the Accuracy dialog box under the World menu. For most purposes you can simply choose the Fast or Accurate mode of simulation, and let Working Model automatically choose a time step and other simulation parameters for your problem.

For those interested in precisely controlling the simulation speed and accuracy, more simulation variables are available when the **W** More or **M** More Choices button is clicked, as shown in Figure A.1.

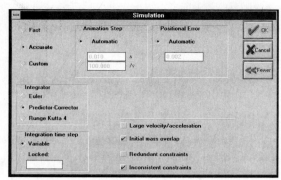

Figure A.I

Fast/Accurate

These buttons set the simulation parameters to the following defaults:

	Fast	Accurate
Integrator	Euler	Predictor-Corrector
Integration time step	Locked	Variable
Warn large velocity / acceleration	Yes	No
Warn initial mass overlap	Yes	Yes
Warn redundant constraints	No	No
Warn inconsistent constraints	No	Yes

Custom

If you change any simulation parameters after selecting either Fast or Accurate, the simulation mode changes to Custom.

Integrators

Working Model assigns an integrator when you choose either Fast or Accurate mode. This section is included for those who wish to experiment with the benefits of various integration methods.

The integrator is the mathematical process that continuously integrates objects' accelerations to update their positions and velocities. The following integrators are available in Working Model:

Euler integration

Predictor-Corrector integration

4th-order Runge-Kutta (Runge-Kutta 4) integration

In order to illustrate the relative complexity of the three methods, let's examine the following first-order differential equation:

$$x = f(x,t)$$

and see how each integrator solves it numerically. You are interested in solving x_{n+1}, or the value of x at the "next" step t_{n+1}, given the information x_n and t_n.

Euler

Euler integration is the fastest and simplest (but least accurate) integrator available for a given time step. Euler integration is the default in the Fast simulation mode and should suffice to give you a rough idea of the motion.

The Euler method solves the preceding differential equation in a single step:

$$\dot{x}_{n+1} = x_n + \Delta t \cdot f(x_n, t_n)$$

Predictor-Corrector

Predictor-Corrector integration is a relatively simple and common scheme for obtaining increased accuracy. For each time step, the integrator computes the states (position and velocity) of objects through two calculations (a two-step method).

The Predictor-Corrector integrator is accurate for common problems such as projectile motion, oscillators, fields, and rolling objects. Predictor-Corrector integration is the default integrator in Accurate mode.

The Predictor-Corrector method solves differential equations in a pair of computations—called predictor and corrector, hence the method's name—as follows:

predictor:

$$x^*_{n+1} = x_n + \Delta t \cdot f(x_n, t_n)$$

corrector:

$$x^*_{n+1} = x_n + \frac{\Delta t}{2} (f(x^*_{n+1}, t_{n+1}) + f(x_n, t_n))$$

where x^*_{n+1} is an intermediate variable used to compute x_{n+1}.

Runge-Kutta 4

This is an the most accurate stable integration method available in Working Model. To use this integrator, you must select the Custom option. At each time step, the Runge-Kutta 4 integrator calculates the accelerations of each object through four calculation steps. The method is considerably slower, but much more robust than the Euler integrator.

Runge-Kutta 4 integrates the differential equation as:

$$x_{n+1} = x_n + \Delta t \left[\frac{1}{6} f(x_n, t_n) + \frac{1}{3} f(x^*_{n+0.5}, t_{n+0.5}) \right.$$
$$+ \frac{1}{3} f(x^{**}_{n+0.5}, t_{n+0.5})$$
$$\left. + \frac{1}{6} f(x^*_{n+1}, t_{n+1}) \right]$$

where

$$x^*_{n+0.5} = x_n + \frac{\Delta t}{2} f(x_n, t_n)$$

$$x^{**}_{n+0.5} = x_n + \frac{\Delta t}{2} f(x^*_{n+0.5}, t_{n+0.5})$$

$$x^*_{n+1} = x_n + \Delta t \cdot f\left(x^{**}_{n+0.5}, t_{n+0.5}\right)$$

Animation Step

The animation step box determines the time between frames of animation. Data from the simu-lation is presented on the screen as a new frame at this interval. This box does not represent the integration time step.

By default, Working Model automatically tries to choose a good animation step, based on the type of simulation that has been created. Using the size of the objects, their spacing, and their velocities, Working Model determines an ideal time step size. Thus, if you are modeling the solar system, its large size, spacing, and velocities force an automatic animation step size in the range of hours or days.

You can override the automatic animation step decision and set your own time step size between animation frames. See the next section, Integration Time Step, for details on how to set the integration time step.

Integration Time Step

The integration time step determines the step size of the numerical integration. See the section Numerical integration on page 260 for more details on the time step.

The integration time step is typically the same size as the animation step, but you can make it less (if less, one animated frame reflects results from several integration time steps).

Locked

In the Fast mode, the integration time step is locked and equal to the animation step. You can make the integration time step smaller than the animation step by typing the desired step size in the text box. By so doing, you are packing multiple integration steps into one animation frame.

Variable

Variable time steps are useful for situations where large accelerations or velocities occur within a small portion of a simulation's duration. In these situations, the integrator reduces the integration time step when accelerations or velocities are high, for better accuracy.

When possible, the integrator increases the integration time step to improve computational performance.

In the variable time step mode, Working Model computes time steps in the following fashion. Upon completing one animation frame, Working Model computes the next frame in two ways:

Using the given animation time step

Using two half animation time steps (therefore computing two small steps)

Before displaying the next frame, Working Model compares the results of the two methods. If the discrepancy between the two exceeds a certain tolerance, the current integration time step (which is equal to the animation step in the beginning) is considered to be too large for the particular frame. In this case, Working Model uses an adaptive refinement method to find a smaller integration time step, taking into account how much discrepancy was encountered previously. Working Model recursively repeats the process until the discrepancy falls within the tolerance, and the remainder of the animation frame is computed using these smaller integration time steps. After the frame is finished, however, Working Model resets the integration time step size to the animation step size and starts all over again to compute the next frame.

This process effectively breaks one animation frame into multiple intermediate frames (which are not displayed on the screen) in order to ensure accuracy. As a result, you may notice that some frames require more time to display than others, since the animation time step remains constant throughout the simulation.

If accelerations become very large during a simulation, the integrator may be unable to find a time step small enough to meet the accuracy criteria. In such cases, a warning is given to indicate that an acceleration has become too large to continue (see the Warnings section).

Positional Error

During the course of a simulation, Working Model is constantly monitoring various types of potential errors, such as:

Interpenetrating objects

Constraint violations

At each integration step, Working Model checks its computation results to see if the model satisfies the error bounds. The value displayed in the Positional Error window is used by Working Model as the basis for determining the tolerable error bounds.

For example, in order to prevent two objects from interpenetrating, Working Model applies a repulsive contact force to each object when the objects overlap by more than a fraction of the value in the Positional Error box. This scheme ensures that the overlap will never exceed the value specified in the Positional Error box. See Simulating Collisions on page 265 for more details.

Similarly, a body moving along a curved slot is subject to this error checking. At each step, deviations of the body from the slot trajectory exceeding the error bounds are corrected by an appropriate force.

Pin joints are more closely monitored for error correction. Maintaining the pin-joint constraint is a fairly simple process, and Working Model corrects errors as soon as they appear.

By default, Working Model automatically computes an appropriate value of Positional Error for a given model, based on the properties of mass and constraint objects therein.

If necessary, you can override the default and specify the value (it must be greater than zero). Keep in mind that as the tolerable error becomes small, Working Model may have to spend more computation time monitoring and preventing errors. On the other hand, an excessive tolerance may produce inaccurate simulation results.

Warnings

When a warning occurs, the simulation pauses and displays a dialog box. You can then stop or continue the simulation.

High velocity/acceleration

Warns when objects have a velocity or an acceleration large enough to violate the tolerance specified in the simulation (see Positional Error on page 264). You can override this warning at run time, but the remainder of the simulation may be inaccurate.

Initial mass overlap

Warns when two or more masses overlap by more than the set tolerance (see Positional Error on page 264) at the initial condition, and the objects are not connected by joints or designated as Do Not Collide.

When two overlapping objects collide, they may cause physical instability in the simulation. See Preventing unstable simulations on page 266 for more information.

Redundant constraints

Warns when there are more constraints than necessary to constrain a specific object's motion. For example, a mass object with several pin joints between it and the background has redundant constraints.

Inconsistent constraints

Warns when incompatible situations occur; for example, when an object driven by a constant velocity motor hits a second, anchored mass. You can override this warning to continue the simulation, but the simulation results may be spurious.

Simulating Collisions

Since Working Model numerically integrates in discrete terms, mass objects can overlap with others by a small amount. For instance, two objects may be close to each other but apart at time step n, but their velocities may cause them to overlap each other in the next frame, at time step $(n + 1)$.

In the variable time step mode, you can define the error tolerance to bound the amount of overlap (see Positional Error on page 264). Working Model automatically uses appropriately small integration steps near collisions and maintains the overlap within the tolerance.

Working Model detects collisions geometrically by finding intersections between objects. When objects are colliding, Working Model computes the forces necessary to prevent interpenetration. Based on these forces, Working Model calculates the new velocities of the bodies and continues the simulation.

Since collisions are simulated in discrete time, the time history of the forces arising from the collision is affected by the time step. In physical experiments, a collision force function typically resembles a spike-shaped bell curve whose support (where the function is non-zero; i.e., the physical duration of the collision) is often much smaller than a typical time step used in numerical simulations. Working Model preserves the area of the function, resulting in peak values that approach experimental values as the step size becomes closer to the physical duration of the collision.

As the time step size becomes smaller, the numerical peak value approaches the experimental peak value.

Simulation Accuracy

Preventing unstable simulations

An unstable simulation is indicated by objects moving in random directions at high velocities. When simulations become unstable, it is immediately apparent.

Simulation instabilities can occur when objects that initially overlap are allowed to collide. Any two mass objects that are not connected with a joint can collide. If you overlap two objects without using the Do Not Collide command, large forces will be generated to move the objects apart.

Instabilities usually indicate the need for a smaller time step. If an object is moving large distances in a short time and interacting with another object through a joint or contact, incorrect results and instabilities can result.

A good rule of thumb is that the time step must be small enough to capture small motions that occur in the system. If you are modeling a guitar string, you will need a very small time step. If the guitar string oscillates 440 times per second, you would need at least four time steps to accurately model each back-and-forth motion of the string. Thus you would need 1760 time steps per second, or a time step of 1/1760 second.

Other systems requiring a small time step include very heavy objects interacting with very light objects, chains that are being stretched, and light wheels on heavy cars.

You can usually correct instabilities with the Accurate simulation method (use the Accuracy dialog box). The Accurate simulation method automatically adjusts the time step for you. As an alternative, use the Fast simulation method, but decrease the time step. As you experiment, you will get a feel for how the time step affects the simulation.

When using the Fast method (fixed time step), reducing the time step increases the accuracy.

When using the Accurate method (variable time step), reducing the value in the Positional Error field increases the accuracy. The Positional Error value determines how much positional error is allowed during each time step. The smaller the error value, the more accurate the simulation.

Getting high accuracy simulation

The method used to increase simulation accuracy is similar to that used with finite element analysis packages. Run your simulation several times at increasing accuracies until the results begin to asymptotically converge.

There are two ways to increase accuracy: With the Fast method (fixed time step), decrease the size of the time step. With the Accurate method (variable time step), decrease the Positional Error value.

The best terms to use as a check for accuracy are the positions and velocities of mass objects that are integral to a system and subject to large velocities. When simulating a vehicle suspension system, use one of the fast-moving suspension components to check for accuracy. When simulating a swinging chain, use one of the outermost links.

To check for accuracy with a specific component, create position and velocity meters for the component. Record the meter values at a specific time near the end of the simulation. Reset the simulation and decrease the time step by half if using a fixed time step method, or decrease the Positional Error value by half if using a variable time step method. Run the simulation again and see if the metered values change significantly. By running with increasing accuracy, you should see any value that you measure converge from simulation to simulation.

If you do not see convergence in a measured simulation variable, there is a good chance that the system is highly dependent on initial conditions or is unstable in some way.

Systems that gain energy

Systems that gain energy usually need to be simulated with a more accurate simulation method. If a simple pendulum begins to swing higher and higher, the system is gaining energy. If a system is gaining energy, reduce the time step or try using a variable time step.

Mass objects and points with equations

Use care when defining the positions of mass objects and points with equations, especially when joints are involved. Mass objects and points that are joined to other objects can cause unreliable results if their position equations are discontinuous.

Split all joints that contain equation-based points, and then click the single-step forward button on the Tape Player controls. This runs one frame of simulation, and loads all equation-based points with their correct positions. Then reset the simulation, and join all the joints.

This prevents points that are defined by equations from jumping on the initial simulation step.

Accuracy and system properties

Simulation accuracy depends to a large extent on the physical system being modeled. Certain physical systems, such as a four-bar linkage with a single driving force, lend themselves to very accurate simulation. Physical systems that have a high dependence on initial conditions lend themselves to simulations that are also sensitive to initial conditions.

If your system would never produce the same results twice in a row in the real world, you can only expect a simulation to give you insight into possible behaviors of the system. Dropping a linked human figure down a flight of stairs is an example of this.

If your system is reproducible in the real world, you should be able to get simulations accurate to any degree you choose.

Why there are numbers like 1e–19 in position fields

These numbers are caused by round-off that occurs when dragging objects with Grid Snap turned on. After a number of drags, very small differences in the last digit of large floating point numbers can accumulate. The number 1e–19 means 0.0000000000000000001. This round-off is so small that it will not affect your work.

If you wish, you can change the numeric display settings in the Numbers & Units dialog box to make this number appear as 0.000. Choose the Fixed decimal point display type. (You can also replace numbers like 1e–19 with the value 0.0 if you wish.)

Precision of meter data in non-SI units

Meters that measure forces, energy, or power can be slightly less accurate in non-SI unit systems than they are in SI. For example, suppose you create a meter to measure the translational kinetic energy of a mass. Working Model uses $\frac{1}{2}mv^2$ as the formula. The SI unit system is designed so that 1 J (Joule) = 1 kg-m^2/s^2. However, in the English unit system, where Btu (British Thermal Unit) is used to measure energy, 1 Btu is not equal to 1 lb-ft^2/s^2. Therefore, if a meter is to report correct values in Btu while using $\frac{1}{2}mv^2$ to compute energy, a special conversion mechanism needs to be used.

As a result of this conversion, the formula in the translational kinetic energy meter appears as:

9.49e–4*(0.5*mass[1].mass*0.454*sqr(mass[1].v* 0.305))

where the constants 9.49e–4, 0.454, and 0.305 convert Joule to Btu, pounds to kilograms, and feet to meters, respectively.

Since these constants are generated with the number of digits specified in the Numbers & Units dialog box, meters may present slight discrepancies between the SI and non-SI unit systems. You can increase the number of sub-decimal digits (up to 6) in the Numbers & Units dialog box to increase the precision of these conversion factors generated by Working Model. If you want more than 6 digits, you can manually type the constants with more digits. Constants in the formulas are stored as typed (as long as the total length of the equation is less than 255 characters).

Tips for Using Working Model

Selecting connected objects

The pop-up menu at the top of the different Utility windows (Properties, Appearance, and Geometry) is a useful way to select objects in your simulation. You will notice that some of the entries (points, masses, and constraints) appear highlighted (**W** preceded by a *, or **M** in bold type). These entries are related in some way to the current selection.

If you select a mass object, all of the points that are connected to the mass appear highlighted on the pop-up menu. If you select a point, the mass object to which the point is connected appears highlighted. If you select a constraint, the points associated with the constraint appear highlighted.

What if a point or constraint won't drag?

The Lock Points feature on the View menu may be turned on. Lock Points prevents points from moving relative to their mass objects. Use the Lock Points feature when editing complex linked figures or mechanisms. Then you will not have to worry about inadvertently changing the geometry or offset of a joint by dragging a point.

Overlapping points show up as not selected

When you split a pin joint, the result is two points that lie directly on top of each other. Each point is connected to a different mass object. If you use a selection rectangle to select the two points, they will appear as if neither is selected. This is because Working Model uses a fast exclusive-or algorithm to draw selection highlighting. You can verify that both points are selected by choosing Properties from the Window menu. The Properties utility window will show "mixed selection" on the top menu.

Using rigid joints to build complex objects

You can use rigid joints to build large, complex objects from simple shapes. It is easier to create a hollow box shape by rigidly joining four rectangles than it is by sketching a complex polygon. Rigid joints do not introduce extra equations of motion into a simulation, and thus they are preferable to using two pin joints when locking objects together.

Settling objects

The Working Model simulation engine can be used to align objects. Take for example a block that needs to rest exactly on an inclined plane. Select both the block and the plane, and set their frictional coefficients to a high value like 1.0. Place the block so it is approximately in position over the plane, and then run the simulation. The block comes to rest in a stable position. Stop the simulation and choose Start Here from the World menu. This makes the stable, settled position the initial conditions. The block is perfectly aligned on the plane.

Placing points directly on the edge of a mass object

To place a point directly on the edge of a mass object, first sketch the point inside the mass: near to, but not on, its edge. Then select the point and choose Properties from the Window menu. In the Properties utility window, enter an offset that will place the point directly on the mass object's edge.

Troubleshooting

The force on a rigid joint measures 0

Make sure the joint is set to Measurable mode (toggle the radio button in the Properties window of the joint).

All points in my DXF file come out attached to background

Working Model is actually designed to behave this way, since DXF™ drawing tells nothing about what points belong to what objects. You can attach individual points to mass objects by selecting the points and objects and choosing Attach to Mass on the Object menu.

My DXF-imported drawing looks strange

Working Model applies a set of well-defined conversion rules when it imports a DXF file because not all the design primitives available in DXF can be interpreted in Working Model. Please refer to Appendix C for detailed descriptions of the steps involved in importing DXF files.

I cannot select points inside mass objects

Point objects are always drawn in the graphics layer below that of mass objects. When the fill-pattern of a mass object is set transparent—such as polygons and circles imported from DXF files—you can still observe the points that are actually covered by the mass objects. But you cannot move the mouse over to those points and click on them.

You can select the points by using box-select. Start the box-select from somewhere outside the mass object covering the points, and draw a box-select rectangle so that the desired points are within the box. No mass object is selected unless it is completely within the boundaries of the box-select rectangle.

Compatibility with Working Model 1.0 files

Working Model Version 2.0 corrected the self-modifying-values problem occasionally seen in Version 1.0. If you come across a problem related to the unit conversion when you open a 1.0 file (for example, the Properties window displays a number different from what you typed), change an arbitrary unit in the Numbers & Units dialog box, and then change it back. This action forces the file to update the unit handler to the one provided in Version 2.0. Save the file again with Version 2.0 so that the file will have the improved unit handler.

w Printing

In Windows, if Working Model holds a large amount of simulation history, printing may result in an error message like "printer overrun." Discarding or making the history smaller will resolve the problem. You can select Start Here on the World menu (please note that the simulation history will be discarded and the current frame will be the initial frame for the next simulation).

Curved slots

When a closed curved slot has uneven distribution of control points, the shape of the slot may appear anomalous, depending on its orientation. Simulations and model editing work correctly and are not affected by this graphic anomaly.

When a curved slot has a tight loop or a sharp corner, the attached object may "jump" from one section of the slot to another while you are running the simulation or dragging the object. You can fix this problem by decreasing the time step and/or Positional Error value in the Accuracy dialog box. Objects never jump between distinct slots.

Working Model may respond rather slowly while you are dragging an object attached to more than one curved slot.

When you use Copy Table in the Geometry utility window with interpolated points, and if two adjacent control points are identical (have the same coordinates), one of the points is not copied to the Clipboard.

DXF import

Working Model limits the number of DXF objects that can be imported at one time to 1500. This limit exists to ensure that Working Model's performance will not be seriously degraded when handling a large number of objects.

Working Model can import two-dimensional objects. All other objects are filtered out at load time. Attempts to import 3D files will have unpredictable results.

While DXF files contain object dimensions, they do not specify a unit system. Before importing a DXF file, make sure that the current unit system in Working Model coincides with the one that you used in your CAD system to create the drawing. When importing DXF files, Working Model automatically assigns its current unit system to the shapes described in the DXF file. (If you create a DXF shape that you meant to be 1 inch wide, and import this shape while Working Model is using meters, the shape will show up as 1 meter wide.)

Each CAD program uses its own rules when converting drawing objects into the DXF format. We suggest you consult the appropriate sections in the CAD manual to understand how the CAD objects are converted to DXF.

Polyline DXF objects with two vertices are converted to lines.

Exporting

When using export options that use a frame-by-frame export (MacroMind Three-D™ or WaveFront™) on a document with a large number of objects, you might experience a delay before the frame progress bar starts moving. This is because the first step of exporting data is to write geometry information, before writing the frame-by-frame position information. The delay will be longer if the system has never been simulated before (i.e., there is no time history present).

W Ropes

Dragging the endpoint of a rope with the Control key does not allow you to change which objects the endpoint is attached to. As a workaround, you can drag the rope over to the target object (without pressing the Control key), and set the desired length of the rope in the Properties window.

constraintforce[ ]

The formula language constraintforce[ ] returns 0 when the constraint is not active. Refrain from defining the Active When field by referring to the force exerted by itself, or your simulation may incur anomalies.

Formula Language Reference

This appendix describes the Working Model formula language. It covers the following subjects:

- Formulas
- Numeric conventions
- Fields
- Operators
- Functions
- Predefined values

Formulas

Working Model formulas follow standard rules of mathematical syntax, and strongly resemble the equations used in spreadsheets and programming languages. Formulas are composed of identifiers, fields, operators, and functions. The following sections discuss each of these categories in detail.

Formulas can be up to 255 characters in length. Capitalization and spacing do not affect formulas, although identifiers and function names must not contain spaces. Parentheses behave as they do in standard algebraic manipulations.

Numeric Conventions

Numbers in Working Model use the standard scientific notation of spreadsheets like Lotus® 1-2-3® and Excel and computer languages such as PASCAL and C.

Exponents

Exponents are displayed in the following way:

In Printed Text	In Working Model
123×10^3	123e3
1.001×10^{-22}	1.001e-22

Angle measures

All angles are expressed in radians. An angle of 360° has a radian measure of 2π. (Note that angles in formulas are expressed in radians, even though the default display mode for Working Model is degrees.)

Identifiers

Identifiers are used in formulas to identify an object. There are five types of identifiers. When creating formulas, you can use one or more of these types:

Mass[3]

Point[2]

Constraint[44]

Output[12]

Input[5]

The number inside the brackets is the object ID. Each object in Working Model has a unique ID. To find the ID of an object, double-click on the object to display its Properties utility window. The ID appears with the proper formula syntax at the top of the Properties window. In addition, the identifier of an object is displayed in the Help Ribbon when the pointer is over the object.

For example, Mass[10] is the ID for mass object #10.

Mass[ ]	Mass[ ] is the identifier for mass objects, such as circles, polygons, and rectangles.
Point[ ]	Point[ ] is the identifier for point objects. Point objects are either isolated points or the points that compose the endpoints of a constraint.
Constraint[ ]	Constraint[ ] is the identifier for constraint objects, including springs, ropes, joints, and pulleys.
Output[ ]	Output[ ] is the identifier for all meters.
Input[ ]	Input[ ] is the identifier for all input controls, including sliders, text boxes, and buttons.

If you use an identifier with an ID for an object that does not exist, the result will be a "Null" object. Null objects return 0.0 for all of their properties.

Fields

Each identifier in the Working Model formula language can have fields. You use fields to access the values of basic properties, such as position and velocity. Fields are a way of accessing smaller components of some bigger object.

Fields are specified by a type, followed by a period (.) and a field name. To access the moment of a mass object with an ID of 3, you would enter the formula:

mass[3].moment

The value returned from mass[3].moment is a number that can be used in any formula.

Any value that has x, y, and rotational components is returned as type vector. The vector type has three fields: .x, .y, and .r (rotation).

Sometimes you will use two fields in a row. To obtain the rotation of a mass object, you enter the following formula:

mass[2].p.r

This equation has two fields. First, mass[2].p produces the position field of mass object #2, which is a vector. Next, .r produces a rotation value from the position field.

mass[2]	mass type
mass[2].p	vector type
mass[2].p.r	number type

Following is a list of all fields with the type of value that each field produces.

Type	Field	Type Returned
Vector	.x	number
	.y	number
	.r	number
Mass	.p	vector
	.v	vector
	.a	vector
	.mass	number
	.moment	number
	.charge	number
	.staticfric	number
	.kineticfric	number
	.elasticity	number
	.cofm	Point
Point	.p	vector
	.v	vector
	.a	vector
Constraint	.length	number
	.dp	vector
	.dv	vector
	.da	vector
Output	.x	number
	.y1	number
	.y2	number
	.y3	number
	.y4	number
Input		number

Vector fields

x, y, r

Notice that position, velocity, and acceleration are always returned as type vector in the preceding table. For example,

point[4].a	acceleration of point #4
mass[3].v	velocity of mass center of mass #3

are both of type vector. You cannot enter these formulas in a text field, because they are not numbers.

To access individual components of these vectors, you must designate whether you want the x, y, or rotational components.

To get a number, enter the following equation instead:

mass[3].v.x   x velocity of mass center of mass #3

mass[3].v.x represents the x-velocity of the mass center of mass #3, which is a number, not a vector.

The subfield .r returns the rotational component of any vector.

Mass fields

p, v, a
These are the current values of position, velocity, and acceleration. Each of these fields returns a value of type vector. Thus, to use any of these fields, you need to add one of the vector fields (x, y, or r).

mass[1].p.x   x position of mass center of mass #1

mass[3].v.y   y velocity of mass center of mass #3

mass[37].a.r   angular (rotational) acceleration of mass #37

mass, moment, charge, staticfric, kineticfric, elasticity
These are the current values of the various properties.

mass[3].charge   charge of mass #3

mass[14].mass   mass of mass #14

cofm
This field returns a value of type point. Thus, the equation:

mass[3].cofm

could be used in any formula requiring a point.

Point fields

p, v, a
These are the current values of position, velocity, and acceleration. Each of these fields returns a value of type vector. Thus, to use any of these fields, you need to add one of the vector fields (x, y, or r).

point[1].p.x   x position of point #1

The position of a point is given in terms of the absolute coordinates.

Constraint fields

length
This is the current distance between the two points of the constraint. To find the current length of a spring, you would enter:

constraint[3].length   length of constraint #3

dp, dv, da
These are the current values for the difference in position, velocity, and acceleration between the two points of the constraint. Each of these fields returns a value of type vector.

These variables use the constraint's reference frame when taking measurements. The x value is measured along the line connecting the two points of a point-to-point constraint.

To find out how fast the length of a spring is changing (the difference in velocity between the two endpoints of the spring), enter the following formula:

constraint[3].dv.x

Output fields

x
This is the value that is displayed on the x axis of an output graph.

output[6].x   value displayed on x axis of output 6

y1, y2, y3, y4
These are the values that are displayed on the y axis of an output graph.

output[6].y1   value displayed on y1 axis of output 6

output[6].y2   value displayed on y2 axis of output 6

output[6].y3   value displayed on y3 axis of output 6

output[6].y4   value displayed on y4 axis of output 6

Operators

Operators include all of the common algebraic symbols (+, -, >, =). The following operators require one or two numbers. The letters *a* and *b* are used as placeholders for any number or formula that evaluates to a number.

Numeric operators

The following is a listing of numeric operators that are available for use in formula entry:

Operator	Input(s)	Output
- (negate)	a	-a
+ (plus)	a + b	a + b
- (minus)	a - b	a - b
* (multiply)	a * b	a x b
/ (divide)	a / b	a / b
% (mod)	a % b	a mod b
^ (power)	a ^ b	a^b
>	a > b	1 or 0
<	a < b	1 or 0
>=	a >= b	1 or 0
<=	a <= b	1 or 0
= (equal)	a = b	1 or 0
<>(not equal)	a <>b	1 or 0

These operators require numbers as their inputs. This means that you cannot add most formula elements that are not numbers.

Incorrect:

mass[3] + point[3]   cannot add a mass to a point

mass[3].p - 34.5   cannot subtract a number from a vector

point[7].v + mass[3]   cannot add a vector to a mass

mass[3].p > 44.0   cannot compare a vector to a number

Correct:

mass[3].p.x + point[3].p.x

mass[3].p.x - 34.5

point[7].v.y - mass[3].v.y

mass[3].p.y > 44.0

mass[3].p.y = 44.0

mass[3].p.y ! = 44.0

- (negate)
Takes a single number and returns its negative.

+ (plus)
Takes two numbers and returns the sum.

- (minus)
Takes two numbers and returns the difference.

* (multiply)
Takes two numbers and returns the product.

/ (divide)
Takes two numbers and returns the quotient.

% (mod)
Takes two numbers and returns the remainder of the first value divided by the second.

^ (power)
Takes two numbers and returns the first value raised to the power of the second value.

> (greater than)
Takes two numbers and returns the value 1 if the first value is greater than the second value. Otherwise, returns the value 0.

< (less than)

Takes two numbers and returns the value 1 if the first value is less than the second value. Otherwise, returns the value 0.

>= (greater than or equal to)

Takes two numbers and returns the value 1 if the first value is greater than or equal to the second value. Otherwise, returns the value 0.

<= (less than or equal to)

Takes two numbers and returns the value 1 if the first value is less than or equal to the second value. Otherwise, returns the value 0.

= (equal)

Takes two numbers and returns the value 1 if the two values are equal. Otherwise, returns the value 0. This operator does not assign any value to the left side of the equation. Mass[3].p.y = 3 returns 1 if mass #3's y position equals 3.0. This formula does not set any values of mass #3's position.

<> (not equal)

Takes two numbers and returns the value 1 if the two values are not equal. Otherwise, returns the value 0.

Precedence

Use parentheses to set the order of equation evaluation. All equations are normally evaluated from left to right. Precedence is given to operators in the following order (operators listed in the same row have equal precedence):

( )	[ ]	.		highest precedence
*	/	^	%	
+	-			
<	<=	>	>=	
=				lowest precedence

Arithmetic operators

Operators with the highest precedence are applied first. For example, the following formula:

$3 + 2 * 4$

is evaluated as 3+(2*4) instead of as (3+2)*4. This is because the multiplication (*) operator has a higher precedence than the addition (+) operator.

Use parentheses to change the order of evaluation, or to assure yourself of the order of evaluation if you're not quite sure of the precedence of various operators. In the above example, you could enter the formula as:

$(3 + 2) * 4$

to force evaluation of the addition before the multiplication.

You can nest parentheses, as in the formula:

$((3 + 2) * 4 + 10) / 2$

Be sure to use parentheses, and not brackets ([]) or braces ({}).

Note on inequalities

Although the inequality operators have the same precedences, the return value of the formula:

if (0 < t <= 1, 50, 100)

is actually equivalent to:

if ((0 < t) <= 1, 50, 100)

because the chain of the binary operators is evaluated from left to right. As a result, the preceding formula always returns 50 regardless of the value of t (since $(0 < t)$ returns 1 or 0, the entire first argument is always 1, or true). If you want the effect of "return 50 when t is between 0 and 1, or else return 100," you should type:

if(and(0 < t, t <= 1), 50, 100)

Please refer to List of functions on pages 277–281 for detailed discussions on each function.

Vector operators

The following operators work on vectors.

Operator	Input(s)	Output
- (negate)	vector	vector
+ (plus)	vector,vector	vector

- (minus)	vector,vector	vector
* (multiply)	number,vector	vector
‖(magnitude)	vector	number

These operators require that their input types match those listed in the previous chart. Vector operators are useful for simplifying formulas. To display a meter showing the distance between two mass objects, you would enter the following formula:

|mass[3].p - mass[2].p|

This formula contains two vector operators. First, the - operator was used to subtract the two positions of the mass objects:

mass[3].p - mass[2].p   result is a vector

The preceding chart indicates that the minus (-) operator can be used on two vectors, and that the result is a vector. The result can then be used with the magnitude (‖) operator to produce a number. The following table shows some common mistakes and corrections.

Incorrect	Correct
mass[2].|a|	|mass[2].a|
|mass[2]|.a	|mass[2].a|
|mass[2].a.x|	abs(mass[2].a.x)
mass[2].a.x+ mass[2].v	Unify both operands to vectors or numbers

- (negate)

Takes a vector quantity and returns the negative of the quantity. The .x, .y, and .r fields of the vector are all negated.

mass[3].p.x	value is 10.0
-mass[3].p.x	value is -10.0
(-mass[3].p).x	value is -10.0

In the last case, the value of mass[3].p is negated as a complete vector.

+ (plus)

Takes two vectors and returns a vector that is the sum. The vector that is returned will have each of its fields (.x, .y, and .r) equal to the sum of the corresponding fields of the two vectors being added.

- (minus)

Takes two vectors and returns a vector that is the difference. The vector that is returned will have each of its fields (.x, .y, and .r) equal to the difference of the corresponding fields of the two vectors being added.

* (multiply)

Takes a vector and a number and returns the scalar product. The vector that is returned will have each of its fields (.x, .y, and .r) equal to the product of the number and the corresponding field of the multiplied vector.

‖ (magnitude)

Takes a vector and returns a number that is the magnitude of the .x and .y fields. The magnitude is equal to the length of a line drawn from (0,0) to the (.x and .y) fields of the vector. The number returned from the magnitude function is equal to:

$$|v| = sqrt(v.x*v.x + v.y*v.y)$$

Functions

Functions take from zero to three arguments, and return a number or vector value. All functions accept their arguments in the form:

function(arg1, arg2...)

There are two kinds of functions available. Math functions perform standard mathematical operations. Simulation functions return information from Working Model simulations.

List of functions

Name	Inputs	Output
abs	number	number
and	number,number	1 or 0
angle	vector	number

Name	Inputs	Output
acos	number	number
asin	number	number
atan	number	number
atan2	number,number	number
ceil	number	number
cos	number	number
exp	number	number
floor	number	number
if	number,number,number	number
ln	number	number
log	number	number
mag	vector	number
max	number,number	number
min	number,number	number
mod	number,number	number
not	number	1 or 0
or	number,number	1 or 0
pi		π
pow	number,number	number
rand		number
sign	number	1 or -1
sin	number	number
sqr	number	number
	vector	number
sqrt	number	number
tan	number	number
vector	number,number	vector

abs(x)

Takes a number and returns its absolute value. Example:

abs(mass[3].p.x)

returns the absolute value of mass #3's x position.

and(x,y)

Logical AND operation. Takes two numbers and returns the value 1 if both numbers are not 0. Otherwise, returns the value 0. Example:

and(time>1, mass[2].v.y>10)

returns the value 1 if *time* is greater than 1 and mass #2's y velocity is greater than 10.

angle(v)

Takes a vector and returns the angle the vector makes with the coordinate plane. For example, if a mass has a velocity of 0 in the x direction, and 10 in the y direction, the mass has a velocity that is in the direction of 90° or $\pi/2$ on the coordinate plane. The formula:

angle(mass[3].v)

would return the value of $\pi/2$.

acos(x)

Takes a number and returns its inverse cosine. Values are returned in the range $[0,\pi]$.

asin(x)

Takes a number and returns its inverse sine. Values are returned in the range $[-\pi/2, \pi/2]$.

atan(x)

Takes a number and returns its inverse tangent. Values are returned in the range $[-\pi/2, \pi/2]$.

atan2(y,x)

Takes two numbers and returns the inverse tangent of y/x. This function is useful because, unlike the atan function, it can generate an angle in the correct quadrant. Values are returned in the range $[-\pi, \pi]$.

ceil(x)

Takes a number and returns the smallest integer that is no smaller than the number.

cos(x)

Takes a number and returns its cosine.

exp(x)

Takes a number and returns its exponential (*e* raised to the value of the number).

floor(x)

Takes a number and returns the largest integer that is no larger than the number.

if(x,y,z)

Takes three numbers. If the value of the first number (*x*) is not equal to 0, returns the value of the second number (*y*). Otherwise, returns the value of the third number (*z*). Example:

if(time>1, 20, 0)

returns the value 20 if *time* is greater than 1; otherwise returns the value 0.

Typically, the first argument of an if function is a relation (such as $x > y$) or a logical operation (such as and(*a*, *b*)). You can write nested if-statements by recursively using other if() functions as their own arguments.

For example, shown below is a somewhat naive C-code segment that returns the maximum of three numbers *a*, *b*, and *c*:

```
{
    if (a > b) {
        if (a > c)
                return a ;
        else
                return c ;
    }
    else {
        if (b > c)
                return b ;
        else
                return c ;
    }
}
```

In the formula language of Working Model, the preceding segment can be translated into a single line, as follows:

if(a>b,if(a>c,a,c),if(b>c,b,c))

ln(x)

Takes a number and returns its natural logarithm.

log(x)

Takes a number and returns its base 10 logarithm.

mag(v)

Takes a vector and returns its magnitude. Result is the same as $|v|$.

max(x,y)

Takes two numbers and returns the larger of the two numbers. Example:

max(mass[1].a.x , mass[2].a.x)

returns the larger x acceleration of either mass #1 or mass #2.

If you wish to find the maximum of the three numbers *a*, *b*, and *c*, you could recursively use the max() functions as follows:

max(max(a,b),c)

min(x,y)

Takes two numbers and returns the smaller of the two. Example:

min(mass[1].v.x, mass[2].v.x)

returns the smaller x velocity of either mass #1 or mass #2.

As in the max() function, you could find the minimum of the three numbers *a*, *b*, and *c* as:

min(min(a,b),c)

mod(x,y)

Takes two numbers and returns the remainder when the first value is divided by the second.

not(x)

Logical NOT operation. Takes a number and returns the value 0 if the number is not 0. Otherwise, returns the value 1.

or(x,y)

Logical OR operation. Takes two numbers and returns the value 1 if at least one of the numbers is not 0. Returns 0 if and only if both numbers are 0. Example:

or(time>1, mass[2].v.r>10)

returns the value 1 if *time* is greater than 1 or mass #2's angular velocity is greater than 10.

pi()

Returns the value of π.

pow(x,y)

Takes two numbers and returns the value of x raised to the power of y; i.e., returns x^y.

rand()

Returns a random value between 0 and 1.

sign(x)

Takes a number and returns the value 1 if the number is greater than or equal to zero. Otherwise, returns the value -1.

sin(x)

Takes a number and returns its sine.

sqr(x)

Takes a number or a vector. If the input is a number, returns the square ($x * x$) of the number. If the input is a vector, returns the sum of the .x field squared and the .y field squared.

sqrt(x)

Takes a number and returns its square root.

tan(x)

Takes a number and returns its tangent.

vector(x,y)

Takes two numbers and returns a vector composed of the two numbers. The first number (x) becomes the .x field of the vector. The second number becomes the .y field of the vector.

Simulation functions

Simulation functions are used to extract data from the simulation. These functions are used in the various meters and vectors of Working Model.

Name	Inputs	Output
constraintforce	number	vector
	number,number	vector
	number,number, number	vector
frame		number
frictionforce	number,number	vector
groupcofm	number	vector
kinetic		number
length	number,number	number
normalforce	number,number	vector
section	number,vector	number

constraintforce(x)

Takes the ID number of a constraint (x) and returns a vector describing the current force being applied by the constraint. To find the compression in a spring, use the formula:

constraintforce(3).x

In point-to-point constraints, the .x component of the force vector is always measured along the line connecting the two endpoints. In constraints that apply a torque, the .r component of the constraint force contains the value of the applied torque.

constraintforce(x,y)

Takes the ID number of a constraint (x) and the ID number of a mass (y). Returns the amount of force being applied by the constraint on the mass as a vector. This function is used by meters that measure gravity, air resistance, electrostatic, and custom force fields. The ID numbers for these four constraints are constant, and are described in the next section. The gravity constraint always uses constraint ID #10002. To measure the force imposed on a mass by the linear gravity constraint, use the formula:

constraintforce(10002, 3).y

In this case, the .y suffix is used to get the value of force in the y (up and down) direction.

constraintforce(x,y,z)

Takes the ID number of a constraint (*x*) and the ID numbers of two masses (*y* and *z*). Returns the amount of force being applied by the constraint between the two masses.

This function only returns values for forces that are applied to each pair of masses (planetary gravity, electrostatics, and custom force fields). The ID numbers for these constraints are constant, and are described in the next section. The gravity constraint always uses constraint ID #10002. To measure the force of gravity between two specific masses in a planetary system, use the formula:

constraintforce(10002,3,5).x

As with point-to-point constraints, the .x value of the vector measures the force applied along the line that connects the center of mass of the two mass objects.

frame()

Returns the current frame number. The initial conditions are defined as being frame zero.

frictionforce(x,y)

Takes the ID numbers of two mass objects (*x* and *y*) and returns the friction force of the first object acting on the second. The value is returned as a vector.

groupcofm(x)

Takes the ID number of a group (*x*) and returns the center of mass of all mass objects in the group.

Currently, the only defined group is group #0, which is the group that contains all mass objects.

kinetic()

Returns the total kinetic energy of all mass objects in the simulation as a number.

length(x,y)

Takes the ID numbers of two masses (*x* and *y*) and returns the length of the line connecting their centers of mass.

normalforce(x,y)

Takes the ID numbers of two mass objects (*x* and *y*) and returns the contact force of the first object acting on the second. The value is returned as a vector.

section(x,v)

Takes the ID number of a mass (*x*) and a vector quantity (*y*). Returns the cross-sectional width of the mass object in the direction of the vector. This function is used by the air resistance force field to approximate the drag on mass objects.

Predefined Values

Variables

There are several predefined variables that you can use in formulas.

Name	Type
time or *t*	number
self	mass
other	mass
ground	mass

time or t

Returns the current time in the simulation; *time* always begins at 0.0 in frame #0.

self

Returns a mass type when placed inside a force field equation. When force field equations are evaluated for each mass in the simulation, *self* assumes the value of the current mass on which the force field is being applied. For example, the equation for a linear gravitational field is:

Fy: -self.mass * 9.81

A force is applied to each mass object in the simulation; *self* assumes the value of the specific mass onto which force is being applied. Thus, in this case, each mass has a force applied equal to -9.81 times its own mass.

When force fields are evaluated for each pair of mass objects (pair-wise fields), the value of *self* assumes the mass of the first mass object in each pair.

other

Returns a mass type when placed inside a force field equation. When force field equations are evaluated for each pair of masses in the simulation (pair-wise fields), *other* assumes the value of the second mass of each pair.

For example, the force field equation for planetary gravity is:

-self.mass * 6.67e-11 / sqr(self.p - other.p) * other.mass

or more commonly:

$$\frac{Gm_1 m_2}{r^2}$$

This equation is applied to each pair of mass objects in the simulation. As the equation is applied to each pair of mass objects, *self* assumes the value of the first mass object and *other* assumes the value of the second.

ground

Returns a mass type for the background. This is essentially a mass at location 0,0 that never moves.

Constants

Four ID numbers are reserved for the global force fields of gravity, electrostatics, air resistance, and the custom force field. You will see these ID numbers in the formulas used in meters to measure forces produced by these constraints. These ID numbers are as follows:

Force Field	Reserved ID
gravity	10002
electrostatics	10004
air resistance	10006
custom force field	10008

If you create a meter to measure the force of gravity on a mass, you will see a formula such as:

constraintforce(10002, 3).y

This formula gives the y component of the force applied by constraint #10002 on mass #3. The value 10002 is automatically inserted in the formula for this meter as the constraint ID for the gravity force.

■ A P P E N D I X   C ■

Exporting and Importing

This appendix describes the Working Model export and import options. It covers the following subjects:

- Working Model export options
- Choosing an export data type
- Steps for export
- Exporting DXF files
- Exporting QuickTime movies (Macintosh only)
- Exporting object motion paths or keyframes (Macintosh only)
- Exporting motion data to animation systems (Macintosh only)
- Exporting PICT and PICT animation (Macintosh only)
- Exporting to MacroMind Three-D (Macintosh only)
- Exporting to WaveFront Technologies Advanced Visualizer (Macintosh only)
- Working Model import options
- Importing CAD geometries as DXF files
- Importing Lincages files (Macintosh only)
- Real-time links with external applications
- Windows and DDE
- Macintosh and Apple Events
- Interface objects
- Remote commands
- Simulation cycles and data exchange
- Connecting the inputs and outputs with the application
- Executing remote commands
- A sketch for designing a control system

Working Model Export Options

Working Model can export data in various formats: DXF, text, and additional graphics and animation formats for the Macintosh.

DXF

You can save Working Model simulations as industry-standard DXF geometry files. The DXF format is popular for transferring data between CAD systems. DXF files do not contain motion information; rather, they describe the shape and relative position of objects.

Meter data

You can export the data from any meter as a tab-delineated text file. You can edit this data with a word processor, spreadsheet, or graphics application.

You can also capture meter data by selecting a meter and then choosing Copy Data from the Edit menu. Data from the meter is copied to the Clipboard. You can then paste the data into any application that supports tab-delineated text.

Ⓜ Macintosh export formats

QuickTime movie
The QuickTime format is a standard for animation on the Macintosh. You can play movies exported from Working Model in any application that supports QuickTime.

PICT
You can save a picture of the Working Model workspace as a single PICT file. You can edit PICT files in any paint or draw program, or paste them directly into documents.

PICT animation
Sequential PICT files are used by some animation programs in place of QuickTime. One PICT file is generated for each exported frame.

DXF animation
Sequential DXF geometry files can be saved for each frame of a simulation.

Object positions
You can export motion data from Working Model simulations in the form of tab-delineated object positions. This format is useful for transferring data to animation programs that support pasted-in keyframes.

MacroMind Three-D™ animation
Working Model will export a complete MacroMind Three-D script and geometry files. You can open these files directly in MacroMind Three-D. All motion data will be placed in keyframes, and all objects can be generated as extruded 3D shapes.

WaveFront Technologies Advanced Visualizer™
Working Model will export a complete WaveFront animation script and object files. You can open these files directly in WaveFront. All motion data will be placed in .mov files, all object data will be exported as extruded 3D shapes in .obj files, and a .set file will be generated to tie the information together.

Choosing an Export Data Type

The Working Model export options can be grouped by the type of data they create.

Simulation data

If you wish to export numerical data from a simulation, export meter data. Numerical data includes anything you can measure in a Working Model simulation, such as the force on a joint or an angular acceleration of an object.

Object geometry

Object geometry is the exact shape and size of objects in your simulation. The best format for transferring this information to another application is the DXF format. Most CAD programs support the DXF format.

Ⓜ Macintosh export data types

Animation

If you wish to capture animation, use the QuickTime or sequential PICT export types. QuickTime animations are exported as a single file, and thus are more convenient than sequential PICT files.

Object motion paths or keyframes

If you are using an animation package that gives you access to numerical keyframe data, you can create realistic animations by using Working Model motion data in your animations. The Object Position export type will create a tab-delineated file of each object's position on a frame-by-frame basis. You can export object position data in either row or column format.

Complete animation data for other applications

Working Model supports complete export of object geometry and motion path data to MacroMind Three-D and WaveFront. When using either of these applications, you do not have to paste columns of keyframe data and associate objects with keyframes.

Complete animation export creates a set of object geometry files, as well as complete keyframe data for all objects. Keyframes are generated for each frame of simulation.

Steps for Export

To export any of the various types of data that Working Model supports, use the following steps:

- Create or open a Working Model simulation.
- Choose **File** menu, **Export...**

A dialog box appears, as shown in Figure C.1.

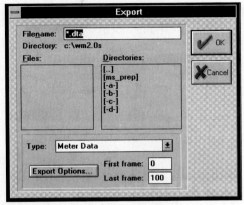

Figure C.1

- Choose the type of data you wish to export by clicking on the menu next to Type.

You will see a list of all the export data types that Working Model supports. Options that are not currently available are dimmed.

- Choose the **Export Options** button to specify particular options for the export data type you will be using.

Each export type has options that are specific to its data type.

- Enter values for the first and last frame.

The current first and last frame of your simulation are placed in the First and Last frame entries. If you haven't run your simulation yet, the last frame defaults to 100. If you have selected a single-frame export type, such as DXF, there will be no last frame and the first frame will be the current frame.

▪ Type a name for the file and then choose **Export**.

A dialog box letting you know that the information is being exported appears on the screen, and the exported data is saved as a file.

Settings from the Working Model workspace apply when exporting. When you are exporting meter data, the numerical format is taken from the current setting in the Numbers & Units dialog box. **M** When you are exporting QuickTime movies and PICT files on the Macintosh, the color settings or number of colors are taken from the current monitor settings.

Exporting DXF Files

The DXF file format can also be used to export the shapes of objects to a CAD or graphics program.

The DXF file created by Working Model is a text file containing an entities section. Working Model objects are converted in the following fashion:

1. Circles are exported as circle entities.

2. Rectangles and polygons are exported as closed polyline entities.

3. Lines are exported as lines.

4. Curved slots are exported as polylines. Open curved slots are exported as open polylines.

5. All other objects are exported as 2D entities.

The current numbers and units settings are used when you are exporting DXF files. If you are using a CAD system that uses inches as the unit of measure, be sure to set the units of your Working Model simulation to inches before exporting.

Special consideration for certain CAD packages

This section covers any considerations that are specific to certain CAD packages.

Vellum   At the time *The Student Edition of Working Model* was released, the latest version of Vellum did not export polylines. Vellum users will have to export the lines from Vellum and select and convert those lines within Working Model.

Exporting QuickTime Movies (Macintosh Only)

QuickTime is the standard animation data format used on the Macintosh. QuickTime files contain sequential images that can be played back as movies in many applications.

Complex Working Model simulations will play back more quickly as QuickTime movies.

Export increment

You can skip simulation frames when exporting QuickTime movies by choosing an export increment greater than 1.

Playback rate

QuickTime movies carry information on how quickly they should be played back. A good starting value is 10 frames per second.

The size of the document window determines the size of the exported QuickTime movie. The settings for your monitor determine the bit depth of the QuickTime movie.

If you have a color Macintosh, setting your monitor's bit depths to 4 in the Control Panel will result in the best performance for Working Model, while maintaining a full range of available colors.

QuickTime movies require a large amount of space on your hard disk, approximately 10 KB per frame. As a result, a 100-frame movie may approach 1000 KB, which is larger than the size of an 800-KB floppy disk. The size of the QuickTime movie is directly related to the size of the image being exported and the monitor's color setting.

You can create QuickTime movies that play simulations in real time by matching the playback rate to the time step in your simulation.

Exporting Object Motion Paths or Keyframes (Macintosh Only)

You can automatically export the positions of all objects in a Working Model simulation. The x, y, and rotational position of each object is stored for each frame of a simulation.

Object position data is exported as tab-delineated text. You can export the positional data as either rows or columns. By default, data is exported as columns. You can change this default with the Export in Rows option in the Export dialog box. If you select Export in Rows, Working Model creates rows of positional data rather than columns. Some animation programs, such as Electric Image™, support pasted-in rows of keyframe data.

Object position data can be used to create keyframes in animation packages. If you are using an animation package that supports pasted-in keyframes, you can build realistic motion based on Working Model simulations.

Exporting Motion Data to Animation Systems (Macintosh Only)

Animation systems such as Electric Image and WaveFront Technologies Advanced Visualizer allow you to define motion paths with rows or columns of positional data. Motion in most 2D and 3D animation systems is defined by a series of keyframes. If an animation is being created that is 100 frames long, several of these frames will exactly specify the key position of objects. The position of objects at other frames will be created by tweening, or interpolating between keyframes.

Working Model simulations specify the exact position of all objects at every frame. When Working Model motion is used in an animation system, every frame becomes a keyframe. At every frame, numbers exactly define the x, y, and rotational position of all objects.

Working Model motion data can be used in any animation package that allows editing of keyframe data. Some animation packages, such as Electric Image, allow you to paste in keyframes as rows of tab-delineated data. Some animation packages allow you to import files that contain motion information, such as WaveFront .mov files. To import motion data to an animation package not directly supported by Working Model, you will have to consult your owner's manual as to the best way of bringing in motion data.

Motion data generated by Working Model is two-dimensional. The Working Model workspace is a place, and all objects are flat. Each object is defined by three positional parameters: x position, y position, and rotation.

3D animation packages can use Working Model motion information. 3D animation packages describe the position of objects with six positional coordinates. Three positional coordinates (x, y, and z) describe an object's location. Three rotational coordinates describe the object's orientation.

When you are exporting motion data into a 3D animation package, three of the six motion coordinates are not used. The x and y positional data from Working Model is used directly in the 3D application. The rotational position data from Working Model is used to define rotation about the z axis in the 3D application.

Exporting PICT and PICT Animation (Macintosh Only)

PICT is the standard graphics format for the exchange of picture data between Macintosh applications. Any paint, draw, or graphics application will open or import PICT files.

Prior to QuickTime, many applications stored animation sequences as a series of individual PICT files. These PICT files are numbered sequentially with the suffix .PICT; for example, Car.PICT 0001, Car.PICT 0002. If you wish to create a series of screen shots, you can use the PICT animation export option to automatically generate a number of PICT files.

Export increment

You can skip simulation frames when exporting sequential PICT files by choosing an export increment greater than 1.

Starting file number

Change this value to start files with a suffix other than .PICT 0001. All sequential PICT files will be placed in the same folder.

Exporting to MacroMind Three-D (Macintosh Only)

Working Model will create complete MacroMind Three-D scripts and 3DGF (shape) files, with all relationships between objects and keyframes. You can open the script directly from MacroMind Three-D.

Working Model creates a script that contains a numerical keyframe for each object in a simulation. Each frame of a Working Model simulation is translated into a MacroMind Three-D keyframe. Motion data from each object is placed in the x-

and y-positional keyframes, and the z axis rotational keyframes. The z positional and the x and y rotational keyframes are set to the value 0.0.

Working Model creates 3DGF files for each object in the simulation. These are placed in the same folder as the motion script, and are correctly referenced by name from the script. Working Model can create flat or extruded 3D shapes when creating the 3DGF files.

Export circles as polygons

Most 3D animation packages support shapes composed of polygon meshes. When exporting a circle, Working Model creates a polygon mesh. You can select how many sides exported circles will have.

Extrude

Click on the box next to Extrude to create three-dimensional objects in the 3DGF files. You can also choose the extrusion depth. For pleasing objects, choose an extrusion depth that is comparable to the width of the smallest object.

Export 3DGF as text files

Use this option if you wish to edit the 3DGF object geometry files by hand.

Exporting to WaveFront Technologies Advanced Visualizer (Macintosh Only)

Working Model will create complete WaveFront motion (.mov) and shape (.obj) files, with all relationships between objects and keyframes stored in a script (.set) file. You can open the script directly from WaveFront.

Working Model creates a script that contains a numerical keyframe for each object in a simulation. Each frame of a Working Model simulation is translated into a WaveFront Three-D keyframe. Motion data from each object is placed in the x and y positional keyframes and the z axis rotational keyframes. The z positional and the x and y rotational keyframes are set to the value 0.0.

Working Model creates .mov files containing positional and rotational data from each object in a simulation. Working Model also creates a geometry (.obj) file for each object in the simulation. These are placed in the same folder as the motion script, and are correctly referenced by name from the script. Working Model can create flat or extruded 3D shapes when creating .obj file.

Working Model Import Options

Working Model can import data in the following formats:

DXF

You can import existing CAD drawings into Working Model as DXF geometry files. The DXF format is popular for transferring data between CAD systems.

Ⓜ Macintosh import formats

Lincages
Lincages is a linkage synthesis package developed at the University of Minnesota and distributed by Knowledge Revolution. You can import mechanisms designed with Lincages into Working Model and set them in motion.

Importing CAD Geometries as DXF Files

The DXF file format was developed by Autodesk to exchange information between AutoCAD and other packages. DXF files contain geometry information for all objects in a CAD drawing.

Working Model directly imports DXF CAD files. Therefore, if you want to simulate a model with object shapes that may be difficult to draw directly in Working Model, or if you have CAD data you have always wanted to see in motion, you can first design your model, using your favorite CAD program, then export the data to a DXF file, and import it into Working Model.

Since CAD packages deal with drawings, and Working Model deals with physical objects, not all the information can be transferred seamlessly. For example, lines in a CAD document do not have a physical equivalent, and consequently they have to be converted to appropriate Working Model objects, such as polygons or curved slots. You can perform these conversions within Working Model.

Incorporating DXF files into Working Model

Typically, you incorporate a CAD drawing into Working Model as follows:

- From your CAD program, save the drawing as a DXF file. (Please refer to Important notes on importing DXF files for preparation of your drawing.)

- Make sure your unit system is consistent with the target DXF file.

- Choose **File** menu, **Import...**

The Import dialog box appears, as shown in Figure C.2.

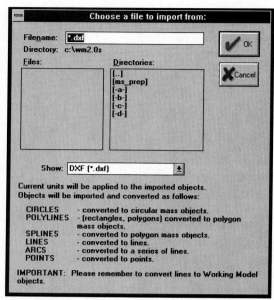

Figure C.2

- Choose **DXF** as the type of file you wish to import.

- Select the file you wish to import.

- Choose **Import**.

The imported objects are placed directly in the workspace. The import process may take some time, depending on the size of the DXF file. You can observe the import process in the dialog box that appears on your screen.

- If necessary, convert selected lines to form polygons, curved slots, and other Working Model objects.

- Attach individual points to appropriate mass objects. Construct joints and other constraints as necessary.

- Assign desired properties (such as mass) to objects. Verify collision specifications between mass objects.

- Run your simulation.

DXF conversion rules in Working Model

Note that a DXF file contains a drawing, whereas objects treated in Working Model are components of a physical model. Therefore, Working Model enforces a set of conversion rules when importing a DXF file. Working Model creates a mass object, a point object, or a line segment for each corresponding entity recognized in a DXF file, and places them in the current workspace, using the current unit system.

After these automatic conversions, you might need to edit the imported model before you run the simulation in Working Model. For example, your original CAD drawing does not contain physical representations of pin joints, slot joints, or springs; in the drawing, they bear only shapes, not meanings. In essence, you must attach physical meanings to what used to be a drawing.

Working Model recognizes the blocks and entities of a DXF file. The conversion rules are as follows:

1. The circle entities are imported as circle mass objects.

2. The polygons and polylines composed of 3 or more vertices are imported as closed, two-dimensional, straight-edged polygons. Polylines composed of only 2 vertices are ignored.

3. Open polylines in the DXF file are translated into closed polygon mass objects.

4. Polylines, splined or curve-fit, are translated into straight-edged polygons. If you wish to import polygons with curved circumferences, we recommend that you approximate the curved portion with a many-faceted, straight-edged polyline in the CAD program before exporting it to the DXF file.

5. Arcs are imported as sets of continuous line segments to approximate the curve. Once arcs are imported as lines, you need to convert them to either polygons or curved slots.

6. Splines are saved in DXF format with the original vertices (control points) and spline-fit vertices. In essence, splines are recorded as a fine-grain polyline, or a polygon of many vertices. Your CAD program may let you decide how many fit points will be computed per spline (AutoCAD, for example, allows you to control this number with the SPLINESEGS command). Working Model will import all these vertices and convert the overall object into a polygon.

7. 3D polylines are read in, but Working Model ignores the third coordinate of each vertex. As a result, Working Model effectively imports a 2D projection of your 3D model. If you wish to import 3D polygons, Knowledge Revolution strongly recommends that you choose a perspective of your 3D model in the CAD program that best illustrates the object *before* exporting it to the DXF file to be sent to Working Model.

8. Lines are imported as lines. However, in Working Model, lines do not have any physical properties; the sole reason for their existence in Working Model is to facilitate the interface with CAD packages. You can resize and move lines in Working Model, but they have no effect on simulations. Once lines are imported, you need to convert them to either polygons or to curved slots.

9. Points are imported as Point objects in Working Model. Initially, all points are attached to the background.

Unit assignment rule

Working Model automatically assigns units to the numbers in DXF files based on the current unit system defined in the Numbers & Units dialog box. For example, if you are importing a DXF file drawn in inches, be sure to set the length units in Working Model to inches before importing.

Important notes on importing DXF files

DXF files created from complex CAD drawing can be extremely large (especially since they are ASCII, not binary, files). Importing large DXF files into Working Model can take a long time (a dialog box will appear on your screen to help you judge how far along the process is). Furthermore, importing DXF files containing hundreds of objects will significantly decrease the speed of Working Model, since it will have to keep track of a very large number of objects.

Knowledge Revolution strongly recommends that you export only the physical objects that you want to simulate from your CAD drawing. Importing a CAD drawing with extraneous lines and objects into Working Model will be slow and time-consuming. The best approach is to edit your drawing in the CAD program before you export it as a DXF file, so that only the objects and constraint attachment points that are relevant to your simulation in Working Model are exported.

Knowledge Revolution also recommends that you use polylines wherever possible when describing rigid-body objects in a CAD drawing that you intend to export to Working Model. Since polylines are automatically converted to polygons at DXF import time, you will benefit from a faster importing process; for example, importing 20 lines takes longer than importing a single polyline consisting of 20 segments.

Converting lines into physical objects

Once they are imported into Working Model, you can convert lines to either polygons or curved slots, using the Object menu.

Attaching points and slots to mass objects

Since a DXF-imported drawing does not contain any information as to which points are attached to which object, you need to specify these relationships in the model with the Object menu, Attach to Mass.

Importing Lincages Files (Macintosh Only)

Working Model directly imports files from the Lincages motion synthesis application. Each link is translated as a polygon. The driving joint is translated as a motor. All other joints are translated as pin joints.

Real Time Links with External Applications

Working Model can exchange data in real time with external applications using **W** Dynamic Data Exchange (DDE) in Windows or **M** Apple Events on the Macintosh. Under this feature, Working Model exchanges data with other applications once every animation time step.

Such links allow you to create a complex control system in another application and drive a Working Model simulation with it. You can also implement complex functions in other applications that may not be supported directly in Working Model. For example, you can implement a look-up table in Microsoft Excel for the horsepower curve of a motor.

Windows and DDE

Working Model can communicate with applications that support DDE (Excel Table or Text for-

mats). These applications include Microsoft Excel, Quattro Pro, MATLAB (version 4.2 or later), and Microsoft Word for Windows. Check with your particular application's user guide to see if DDE is supported.

Macintosh and Apple Events

Working Model can communicate with applications that support the Table Suite of Apple events. These applications include Microsoft Excel and Claris FileMaker. Check with your particular application's user guide to see if the Table Suite is supported.

Interface Objects

Working Model communicates with external applications through meters and controls. Meters can function as output devices, while controls serve as inputs.

Remote Commands

Working Model also allows you to specify commands of external applications during the simulation cycles. You can specify:

Initialize commands, which are executed at the beginning of a simulation

Execute commands, which are executed at every frame during the simulation

Simulation Cycles and Data Exchange

Data exchange and remote command executions are interleaved with simulation cycles of Working Model in the following fashion:

Initialize remote commands

loop while simulation continues {

    get input data from external application

    run simulation step

    send output to external application

    Execute remote commands

}   end loop

Data exchange through **W** DDE or **M** Apple events

Setting up an application interface

To exchange data with an external application:

- Create or open a Working Model simulation.

- Create meters and/or controls for the properties you wish to exchange with the external application.

- Select **Define** menu, **External Application Interface**.

A black interface icon appears in your Working Model document.

- Double-click on the interface icon.

The Properties window for the external document opens in your workspace.

W   Click the Application button.

W   Select the application name from the dialog box.

W   Choose **OK**.

M   Make sure that the application is already running and has the document open.

- Click the Document button.

- Select the name of the document you wish to link to Working Model.

W   Click **OK**.

W The application is launched automatically if it is not already running.

> **W** **Note:** If you save the simulation file with DDE links and reopen it later, Working Model will not launch the application automatically; you must either launch it yourself or retype the document name and press Enter or choose OK.

M   Click **Make Link**.

Connecting the Inputs and Outputs with the Application

At this point, you have specified the external application and the document with which Working Model will interact. You must also specify which individual controls (inputs) and meters (outputs) in the Working Model document correspond to appropriate elements in the external application. For example, you must establish a logical link between controls and meters and particular cells (in Excel) or variables (in MATLAB).

- In the Properties window for the interface object, select an input or output from the list.

- For each selected meter or control object, type in the variable name appropriate for the external application.

For example, if you are using Excel, you can specify the cell by typing:

R1C3

to indicate the cell located at row 1, column 3. In MATLAB, you can simply type the name of the variable exactly as it appears in MATLAB; for example:

x_initial

For meters, observe that all the meter fields (y1, y2, y3, etc.) appear separately. You can specify different variable names or cell names for all these output changes individually.

- Click the **Connect** radio button for the particular input or output.

By default, all inputs and outputs are disconnected. Working Model does not establish the connection with the cell or variable unless you choose Connect. Again, you can connect or disconnect each input and output individually.

- Repeat these steps for all the inputs and outputs for which you want to exchange data with the external application.

- If desired, specify the Initialize and Execute commands appropriate for the external application. See Executing Remote Commands for examples of such commands.

- Run the simulation.
If the external application is used as an output of a Working Model meter object, data is sent to the external application at every frame of the simulation.

If the external application is used as an input to a Working Model control object, data is retrieved from the external application at every frame.

Executing Remote Commands

You can execute commands that are specific to the external application linked to Working Model. For example, you can have MATLAB commands or Excel macros by typing them in Initialize and Execute text boxes before you start a simulation. The following are example commands.

In MATLAB, you can type the function calls into Initialize and Execute command boxes:

(Initialize)   u = 0;

(Execute)   u = f(x,y);

In Excel, you can type Excel macro language into the command boxes. **w** In Windows, the commands *must* be enclosed by a pair of box brackets ([ ]), as shown below (*the brackets are not necessary on the Macintosh*):

(Initialize)   [FORMULA("=R[-1]C+R2C3")]

(Execute)   [RUN("MACRO1")]

where MACRO1 is the name of a macro command you recorded, for example. Please consult the Excel *Function Reference* manual for details.

Commands typed in the Initialize box will be executed when the simulation starts, and before Working Model performs any computation. For example, the commands can initialize data before you start an experiment.

Commands typed in the Execute box will be executed at every frame of the Working Model experiment.

A Sketch for Designing a Control System

As an example, you can use Excel to implement a feedback system that changes the magnitude of the torque of a motor based on its rotational speed.

1. Create an Excel spreadsheet and write a function that describes the feedback control.

2. Make sure a motor in your model has a meter for the rotational speed and a control for the torque. The meter and control serve as the input and the output for the control system, respectively.

3. In Working Model, select External Application Interface from the Define menu. Choose the Excel document. (See Setting up an application interface on page 293 for an explanation of how to define the interface.)

4. Select the motor's control from the lists of inputs shown in the Properties window of the interface.

5. Type the appropriate Excel cell in the Variable field of the output. Make sure the cell contains the desired control function in Excel.

6. Select the motor's meter from the lists of outputs shown in the Properties window. Type the appropriate cell in the variable field. Make sure that the cell is used as the input in the Excel function.

7. Run your simulation.

At every frame, velocity data will be sent to the spreadsheet, whose macro will calculate the desired torque. The torque is returned to the Working Model simulation as a motor input through the control.

Index

A

Acceleration, angular, 94
Acceleration actuators, 70–71
Acceleration motors, 138
Accessing values, 273
Accuracy, 266–268
    checking for, 266
    increasing, 266
    integration and, 141–143
    precise controlling of,
        261–262
    real-world reproducibility
        and, 267
    speed versus, 141–142
    system properties and, 267
    time step and, 261
Accuracy dialog box, GS-32, 61,
        109, 142, 261–262
    correcting instabilities with,
        266
    Positional Error value, 269
Actuators
    defined, 70
    types of, 70
Actuator tool, GS-27, 70–71
Air Resistance, 16–17
    changing, GS-31
    reserved ID, 282
    Standard, 16–17
Alert boxes, GS-41–42
Aligning objects, 268
Anchors
    on ground rectangles, 28
    Smart Editor and, GS-44
    using, 104
Anchor tool, GS-23, 28–29, 104
Angle measures, numeric con-
        ventions for, 272
Angular acceleration, 94
Angular momentum, 149

Angular velocity, 69–70, 214,
        215, 216, 217
Animation data
    exporting, 284
Animation files, exporting, 285
Animation time step, 61, 86,
        263, 264
Appearance window, 24–25,
        74–76, 139–140
    displaying mass centers with,
        226–227
    Show center of mass option,
        83–84
Arithmetic operators, 276
Arrow icon, 28
Arrow tool, GS-21, GS-22

B

Backup procedures
    for Macintosh, GS-9–10
    for Windows, GS-8–9
Batting machine prototype,
        241–252
    running simulation, 248–251

C

CAD files, importing as DXF
        files, 289–290
CAD packages
    transferring between, 284
    Vellum, 286
Cam-followers, 187–198
    running simulation, 193–195
Cams, 188
CAM1.XLS or CAM1, 189, 191
CAM2.XLS or CAM2, 197
CAM3.XLS or CAM3, 197
Center of mass. *See* Mass centers
Central force problems, 64
Chain drives, 217

Charges, 246
Check boxes, GS-41
Circles
    creating, 7–8
    exporting as polygons, 288
Circle tool, GS-23
Circular orbit, 58–59
Clipboard, GS-30
Closed Curve Slot tool, GS-25,
        190–191
Coefficient of restitution,
        changing, 107–108
Collisions, simulating, 265
Colors of objects, changing,
        139–140
Command buttons, GS-40
Commands, choosing from
        menus, GS-28
Compatibility with Working
        Model 1.0 files, 269
Compound gear train, 214
Conditional statements,
        214–215
Constants, 282
Constraintforce[], 270
Constraint[] identifier, 9, 73, 272
Constraints
    defined, 7
    fields, with types of values
        returned, 273, 274
    inconsistent
        message regarding, 212,
            214
        warning about, 265
    redundant, warning about,
        265
    rotational, 138
    Smart Editor and, GS-42,
        44–46, 175–176
    troubleshooting, 270

Macintosh, 258
Windows, 258
Menu bar, GS-287
Menu buttons, creating, GS-36,
    88–90
Menu commands, GS-27–28.
    See also Tool bar
Menus, GS-27–39
    choosing commands from,
        GS-28
    pull-down, GS-40
Metafile data, 164
Meters, 12. *See also* Output
        meters; Velocity meters
    changing scale of, 251
    contact force, 251
    creating, 73, 129–130
    equations defining output,
        129
    exporting data from,
        144–147, 284
    graph display, 109–112
        changing scale, 110–112
    identifier, 272
    measurement in non-SI
        units, 267–268
Microsoft Excel, 189, 191, 292,
    293, 294
    feedback system, 294–295
Microsoft Windows, 292
    backup procedures, GS-8–9
    bracketing of remote com-
        mands, 294
    Help window, GS-39
    installing Working Model in,
        GS-10–12
    mouse used with, GS-6
    requirements for running
        Working Model on,
        GS-5
    RAM requirements, 258
    speed performance, 259
Moment of inertia, 246–247
Motion. *See* Projectile motion
Motion paths, exporting

(Macintosh), 287
Motors, 138–139
    acceleration, 138
    adding, 206–207
    defining properties of,
        138–141
    rotation, 138
    torque, 138
    types of, 138
    velocity of, 138, 207, 224
Motor tool, GS-27, 138–139, 224
Mouse, GS-20
    terminology, GS-6, 7
    using, GS-42, 44–45

N
New Menu Button dialog box,
    GS-36, 88–90
"Null" objects, 272
Numbers and Units dialog box,
    99–100
    changing numeric display
        with, 267
    Distance pull-down menu,
        100, 223
    Energy pull-down menu, 100
    Rotation pull-down menu,
        223, 225
    Time pull-down menu, 223
    Unit Systems pull-down
        menu, 99–100
Numerical data, exporting, 284
Numerical methods, 259–261
    analytical method, 260
    numerical integration,
        260–261
Numeric conventions, 272
Numeric display settings
    changing, 267
Numeric operators, 275–276

O
Object geometry
    defined, 285
    exporting, 285

Object menu, GS-34–35
    Attach Picture option, GS-35
    Attach to Mass option, 269
    Collide option, GS-35
    Do Not Collide option,
        GS-35, 87, 259
    Font option, GS-35
    Join command, GS-34
    Move to Front option, GS-35
    Send to Back option, GS-35,
        193
    Size option, GS-357
    Split option, GS-34
Object position data, exporting
        (Macintosh), 287
Object Position export type, 285
Objects. *See also* Mass objects;
        Point objects
    Appearance window for,
        GS-38, 24–25, 139–140
    changing properties of,
        GS-38
    colors and patterns of,
        139–140
    connectivities of, 73
    controlling, GS-34–35
    converting, GS-35
    duplicating, 137
    Geometry window for, GS-38
    groups of
        changing properties of,
            25–27
    identifiers, 9–10
    joining, 91
    labeling, 24
    moving in workspace, 25–27
    "null," 272
    numbers assigned to, 73
    overlapping, 108
    Properties window for, GS-38
    resizing, 23
    rotating, 181–182
    selecting and deselecting,
        23–24
    splitting, 90–91